NEW VISIONS IN PLANT SCIENCE

Edited by **Özge Çelik**

New Visions in Plant Science
http://dx.doi.org/10.5772/intechopen.72517
Edited by Özge Çelik

Contributors

Lakesh Sharma, Sukhwinder K. Bali, Ahmed Asaad Zaeen, James Dwyer, Luca Marchiol, Phenny Mwaanga, Zhong Chen, An Yan, Kok-Song Lai, Lee Yoon Low, Shun Kai Yang, Janna Ong Abdullah, Kok Andrew-De-Xian, Ngai Paing Tan, Roel C. Rabara, Marilyn C. Ferrer, Jennifer Jara-Rabara, Rachel C. Sotto, Anirudh Kumar, Kapil Sharma, Supriya Sharma, Abhishek Bohra, Abhijit Mitra, Naveen K. Sharma, Özge Çelik

First published in London, United Kingdom, 2018 by IntechOpen
IntechOpen is the global imprint of INTECHOPEN LIMITED, registered in England and Wales, registration number: 11086078, The Shard, 25th floor, 32 London Bridge Street
London, SE19SG – United Kingdom
Printed in Croatia

British Library Cataloguing-in-Publication Data
A catalogue record for this book is available from the British Library

Additional hard copies can be obtained from orders@intechopen.com

New Visions in Plant Science, Edited by Özge Çelik
p. cm.
Print ISBN 978-1-78923-702-3
Online ISBN 978-1-78923-703-0

For EU product safety concerns:
IN TECH d.o.o., Prolaz Marije Krucifikse Kozulić 3, 51000 Rijeka, Croatia,
info@intechopen.com or visit our website at intechopen.com.

Meet the editor

Dr. Özge Çelik completed her MSc on Pharmaceutical Biotechnology at Marmara University and her PhD on Plant Biotechnology at Istanbul University in 2010. She has published more than 20 research papers in national and international peer-reviewed journals along with 2 book chapters and 2 books. She has been working on plant biotechnology for 16 years. Dr. Çelik focused her research on the development of abiotic stress-tolerant plants via mutation breeding and deep molecular analyses of the mutants. She also manages national-funded research projects in relation to research fields. She is an experienced reviewer in many Science Citation Index and non-Science Citation Index peer-reviewed journals. She is presently working as an associate professor at Istanbul Kültür University, Turkey.

Contents

Preface

Plants are important organisms providing food for humans and animals. They are the basis of life on earth. There is a clear progression in plant science. Plant science has several disciplines working together to understand the structural, functional, and evolutional features of plants. Plant species have been adapted to different environmental conditions and respond to various biotic factors to survive. During these adaptive processes, their genetic systems have been shaped and rearranged. Changing environmental conditions and a rapidly increasing world population have further increased the importance of developing new varieties in plant production and hence plant breeding.

In accordance with the latest improvements in molecular biology techniques, transgenic applications, nanotechnology, and omic technologies, it became necessary to discuss applications on these aspects. In this book, we present new technologies having applications on breeding and molecular genetic analysis of plants. Each chapter summarizes current knowledge on different fields of new-generation molecular techniques.

This book is organized into three sections. Section I, Molecular Plant Breeding, describes the molecular marker technologies and transgenic approaches in breeding studies. Section II, Plant Omics, shows the importance and usage of transcriptomic, proteomic, and metabolomic analysis in plant science technologies. Section III, Nanotechnology in Agriculture, sheds light on the use of nanotechnology in agricultural applications.

After commercialization of transgenic plants, biotechnological improvements have progressed. Technological improvements produce genome-scale data to use in breeding studies. The use and processing of plant genome-related data have led to significant improvements in terms of gene expression and the identification of its relationship. Integration of molecular marker analysis systems into marker-assisted selection applications has evolutionary importance in agriculture. The rapid accumulation of sequence resources guarantees that genetic applications will progress with comparative genomics. In this sense, omic technologies have become important. Also, improvements in analytical methods and analysis of metabolites create an insight into the responses of plants against several stress factors. Multi-omics-based systems help to understand the pathways or molecules that have a role in certain plant functions. Genomic information obtained by next-generation sequencing techniques needs to be organized and analyzed. The integrative improvements of multiple omic technologies and computational tools will be helpful in plant biotechnology studies. Besides computational techniques, nanotechnology is a new research area for agriculture. Usage of nanotechnological products as pesticides is gaining favor for new studies.

I am extremely thankful to the contributors for sharing their scientific experiences and knowledge by contributing chapters to this book. I want to express my appreciation to Ms. Marina Dusevic and other officials for their constant help and guidance and for giving me the opportunity to edit this book.

I hope this book will be helpful to a wider reader group, including researchers, scientists, students, and related professionals.

Özge Çelik
Istanbul Kültür University
Faculty of Science and Letters
Department of Molecular Biology and Genetics
Ataköy, Istanbul, Turkey

Section 1

Molecular Plant Breeding

Chapter 1

Introductory Chapter: New Age Molecular Techniques in Plant Science

Özge Çelik

Additional information is available at the end of the chapter

http://dx.doi.org/10.5772/intechopen.79360

1. Introduction

Plants are valuable research objects. Their life spans can reach up to 5000 years, and their survival potential under extreme conditions makes them more interesting. Knowledge about plants has maturated and deeper research areas have been generated.

After commercialization of first transgenic plant, agricultural revolution has been started. Biotechnological improvements have been rapidly integrated into agricultural technologies in response to the global needs.

Plant breeding is an application that changes the plant genetics to thousands of genes, crossing varieties and then selecting the new varieties (and genes) that are desired. In this intertwined event, the plant breeder crosses to ensure that the desired traits are gathered in sufficient numbers, taking into account the preferences for genetic backgrounds. Breeding studies are based on Mendelian genetics. There are several breeding objectives for each cultured plant species. These objectives are possible to alter suddenly. Therefore, new breeding programs should be adapted. At this point, the breeders must be in a close relation with the market and agricultural technologies. Breeding can be described as the continuous period of mating and selection. The only difference is the breeding methods preferred by the breeders.

In practice, biotechnology is often combined with plant breeding to develop plants. In this context, genetic markers mapped near genes responsible for important agricultural features are used to select the desired plant. New age molecular techniques can be easily adapted for plant system. Therefore, a new wide door opened to new possibilities for discoveries in plants.

Releasing the data of plant genomes, it is important to determine the relation between interconnected network of genes and gene products. The requirements of new approaches for analysis and interpretation of the results cannot be denied.

2. Molecular marker analysis

Molecular markers can be expressed as a DNA sequence or gene expression product that represents differences in genomic level in relation to a gene or a property. Molecular markers are markers that can be used to monitor differences at the DNA level and for a gene that is being investigated. DNA markers are also DNA regions in which polymorphism in individuals within a species can be determined [1, 2] .

Molecular markers are nontissue-specific DNA regions that are reliable, repeatable, standardizable, capable of identifying multiple regions in the genome, capable of identifying more than one region in the genome, independent of environmental conditions, dominant and codominant [2–4].

Molecular markers are classified as dominant and codominant markers. Heterozygous individuals cannot be distinguished from homozygous dominant individuals, since dominant markers are not suitable for identification of heterozygous individuals when related to dominance between alleles is dominated by dominant markers. Thus, three different individuals (AA, AA and AA) can be distinguished for any marker at any point [2, 4].

The use of molecular marker systems based on this meta-analysis has become more prevalent in genetic studies conducted by the discovery of the polymerase chain reaction (PCR). The rapid development of technology and the accompanying needs, the facilities of the laboratories where the applications will occur, the biological properties of the species and the abundance of markers in the genome have contributed to the development of DNA markers [5]. Restriction fragment length polymorphism (RFLP), randomly amplified polymorphic DNA markers (RAPD), amplified fragment length polymorphism (AFLP), sequence labeled sequences (STS), microsatellites (SSR), cleaved polymorphic sequence (CAPS), single strand such as complementary polymorphism (SSCP), amplicon length polymorphism (ALP), interspecific sequence repeat polymorphism (ISSR), expressed sequence tags (EST) and single nucleotide polymorphism (SNP) [1, 6–11].

Molecular marker techniques have some advantages and disadvantages as compared to each other, but their reasons for preference vary according to the purpose of the study and the technical possibilities of the laboratory in which the study will be conducted. SSR and SNP markers are frequently preferred because of the high level of polymorphism nowadays [11, 12].

3. Marker-assisted selection (MAS)

Marker-assisted selection (MAS) accelerates the process of developing a new improved variety. Instead of choosing a character (which is a result of many genetic studies), the MAS is based on the genes that provide the desired character. This is called the quantitative feature locus (QTL) the choice of specific alleles in the marker locus is dependent [13, 14]. To summarize the theoretical advantages of MAS:

1. Avoid errors that are caused by environmental changes.
2. Applicability in a juvenile phase without leaving a character.
3. In order to be effective, a single plant may be applied here, while the phenotypic selection of some characters requires many seeds or tissues.
4. Phenotypic selection may be more economical. Although the MAS paternal choice does not take the place of sexual recombination and breeding strategies, it can greatly increase the selection effect of a superior genotype.

Therefore, MAS is considered to be an important technique for improving general plant regeneration. The advantages of MAS may not always be meaningful, and it is often discussed that phenotypic selection for many characters is faster and cheaper than MAS. Some of the factors that may affect the success of MAS in the negative are as follows:

1. Some breeding facilities are inadequate in terms of technical infrastructure and expertise required for the implementation of MAS.
2. Decreasing the influence of the MAS between the marker and the target QTL.
3. Marker must be polymorphic on parents.
4. MAS is only effective if the selected alleles are more important than the other alleles in the population. This last factor is the key to success or failure of every MAS application [13, 15, 16].

As can be seen clearly, MAS is based on the ability to predict the value of alleles. This prediction depends on a number of factors, but it is essentially an allele, the behavior of other alleles in existence and other physical environmental conditions that have not yet been tested. For example, a breeder may determine that the A1 allele at its locus is a positive effect on yield. However, this prediction is made in a limited environment and with a limited number of gene sources. A breeder who crosses a parent with allele A1 and a new parent with the allele A4 and selects the allele A1 as the bound marker will never know that the allele A4 is better than the allele A1, but not in the absence of the allele A1, plants may be susceptible to a disease. For these reasons, MAS should never be applied separately from phenotypic selection. The most successful applications of MAS arise in situations where it is used to improve it rather than applied to phenotype selection [5, 13–15].

4. Genetic linkage mapping

DNA marker technology is used in herbal organisms to study diversity and kinship levels, fingerprint studies, genomic and physical maps, genomic regions associated with various stress factors, and genomic information. Fingerprint analysis aims to identify similarities or differences among genetic materials. Based on the assumption that the variation in genetic markers represents a variation in genes, the use of markers in fingerprint analysis has been conceived. In the fingerprint analysis, markers are widely used to provide information on many of the genomes at the same time. As a result, the proportion of loci that differs between genetic materials is determined. This type of analysis is used to select materials to be imported into the plant rehabilitation program, and with the use of lines with a high variation, the breeder has the choice of choosing what he wants from a wider variation. Fingerprint analysis is also used for various diagnoses. Fingerprint analysis based on genetic markers also has a widespread use in forensic envy. Genetic markers are used in genomic analysis, in evolutionary development, in the identification of structural changes in chromosomes, in genetic resources and in the protection of varieties and in genetic variation. DNA markers are the most trusted and preferred systems because they are not affected by any condition, and because they allow the whole of the genome to be narrated or done [17, 18].

Link maps can determine the position and genetic distance between the markers along the chromosome. A genetic linkage map is formed by determining how often the marker moves together. A good genetic map has many markers on the whole genome without big gaps. The rate of production of genetic maps increased as the rate of use of this information in plant breeding programs increased. Both simple and complex inherited genes can be easily identified by DNA markers [1]. Many characters (such as resistance to certain diseases) that are simply governed by a single gene have been transferred to different genotypes in a very short time, thanks to DNA markers provided that genetic maps are first made. The most effective use of molecular markers has been the refinement of quantitative characters possessing complex inheritance and governed by multiple genes. Many characters such as plant height, flowering time, brooding, yield and yield elements, quality, endurance against certain diseases and harms are being quantitatively controlled and have considerable prospects for plant breeding trials. Since quantitative characters are governed by multiple loci (QTLs), the degree of effect of each locus is different, and because they are highly affected by environmental conditions, it is difficult to determine and transfer in traditional breeding trials [19, 20]. However, due to detailed genetic maps made with molecular markers, the degree of effect of each locus can be determined by locating homozygous populations in different environmental conditions, and probable locations of these loci have been identified on chromosomes. The most important use of link maps is to identify chromosome regions that contain the locus of interest and the quantitative feature locus associated with the feature of interest. These types of maps are called QTL maps or genetic maps. The QTL mapping is based on the presence of markers and genes that open up through chromosomal recombination during meiosis and allow them to analyze this expansion in their offspring [12, 21–23].

Generally, the rate of polymorphism in plant species that can be tolerated is higher than that of self-fertilized plant species. For this reason, partly distantly related rootstocks/parents are selected in the mapping studies carried out on self-fertilized plants [19]. The choice of DNA markers to be used in a mapping study depends on the availability of the currently existing and characterized markers or the suitability of the specific markers for the organism being studied. When polymorphic markers are identified between parental/parent, these markers need to be screened in the entire mapping population. This process is called as marker genotyping [14, 20].

Link analysis can be done manually for several markers, but the use of computer programs is required to perform link analysis for a large number of markers. When genetic maps were constructed to cover a large number of plant species, researchers believed that the genes could be in similar order and in similar sequences in close-up car species. This observation called genetic and collinearity in terms of chromosome organization among species reveals the existence of hundreds or even thousands of common molecular markers that could be genetically mapped in different species. The use of co-markers in mapping studies allows genome-wide comparative analysis of different species [23].

Most DNA markers are selected from nonrepetitive regions in the genome. This means that repetitive DNA is included in the genetic markers as empty and large regions. Along with not being observed much in dicotyledonous plants, while high-order cholinergic activity is observed in monocotyledons, it is observed among some species of synthetic dicotyledonous plants as well as the reason. The strain between species reveals a number of meaningful results. Simply, the genetic information obtained for a species can be transferred to another species by eliminating experimental barriers [14, 24].

The rapid accumulation of sequence resources guarantees that genetic applications will progress with comparative genomics. The linkage of these genomic sources with close relatives and even farther relative species greatly facilitates the exploration of evolutionary narratives. This clarifies the exploration and exploration of important orthologous loci, the restructuring of phylogeny and other biological questions.

5. Omic technologies

The omic technologies makes the interactions understandable between the genes, proteins and the biochemical pathways by using several molecular and analytical methods such as bioinformatics and computational analysis methods. The main focus of omic technologies is the key traits of interest known as genomics, transcriptomics and proteomics. Improvements in instrumentation and analytical methods have driven the major data sources of omic technologies such as genomics, transcriptomics and proteomics forward [25, 26].

Technological improvements produce genome-scale data to use in breeding studies. In relation to the improvements in analytical methods, analysis of the metabolites becomes important. Profiling of the alterations of the metabolite accumulation provides an insight into the responses of the plants against several stress factors. A new omics research field "metabolomics" was born. Nontarget metabolome analysis is also useful to evaluate the tissue specific metabolites and secondary metabolites. It has been reported that significant progresses in metabolite quantitative trait locus (mQTL) analysis have been used in several plant species [27, 28].

The main issue in omic technologies is to combine the heterogeneous data sets. High-throughput quantitative omic data are the best option to describe the different levels of the information of a biological system. Computational tools are effective to overcome this problem. An integrative analysis of the genome-scale data, comparative analysis of the genomes, phytochemicals, and biosynthetic pathways can be easily and successfully performed. Multi-omics-based systems are demonstrative to understand the pathways or molecules having role in certain plant functions [29, 30].

Epigenomics is one of the latest tools to understand a gene function regulation in an organism. The newest technologies have opportunity to enable the data to resolve the mechanisms. Epigenomics provides us ability to define phenotypic variations via DNA-protein interactions, chromatin modifications and RNA technologies. Also, usage of chromatin immunoprecipitation (ChIP) techniques with next generation sequencing (NGS) technologies can gain epigenomic data from plant species [31–33].

6. Next-generation sequencing (NGS)

The improvements in sequencing technologies, an important era in plant genomic researches have been started. In a short time, cost-effective sequencing technologies have been developed. Next-generation sequencing (NGS) platforms give opportunity to plant genomic studies for several breeding strategies [34–36]. It is available to work with the plant genome and the whole transcriptome by using NGS platforms without resequencing. HeliScope™, SMRT™, RNAP™ and Nanopore DNA sequencer are classified as 3rd generation sequencing technologies. Recent advances in DNA sequencing technologies produce new analyze methods to define the exact mechanisms of the traits. Genome-wide association studies known as GWAS are effective to discriminate the complex features in plants. GWAS can scan the molecular markers among the DNA, gene or genome rapidly, and it can be possible to find the genetic variation which is in relation to an agronomic trait. GWAS uses the NGS data to find genetic variations [30, 37, 38].

NGS technology is also effective for characterization of transgene constructs such as flanking regions and other element combinations [39]. NGS technology is more sensitive than qPCR GMO detection to find out the existence of unknown GMOs. Integration of NGS to other new age molecular methods such as DNA walking opens a new window in GMO screening [30, 39–42].

7. Bioinformatic analyses

Genomic information obtained by new-age molecular biology techniques is required to be stored, organized and analyzed. Bioinformatic methods have progressed rapidly and exchanged the status of the research. The use of bioinformatics tools is crucial for the processing of large-scale data in detail.

The important point is to process and analyze plant genomics data. NGS technology is the main challenge. In recent years, the increase in the number of sequenced plant genomes and the need for tools are obvious. The heterogeneous nature of the plants and innovative bioinformatic tools have become mandatory. The German Federal ex situ Gene Bank of Agricultural and Horticultural Crops (GCBN), GIBS (Genebank information system), EURISCO, LAILAPS, PGP&e!DAL, PlantsDB, IPK blast server, Plabi PD are recent platforms for plant genomic resources [43].

The integrative improvements of multiple omic technologies and computational tools are helpful in plant biotechnology studies. Interdisciplinary collaborations are important to enable the network between different fields of life sciences. This must be the most important mission for the researchers working on plant biotechnology to provide new insights on agricultural problems. Otherwise, it will be a big challenge to solve the upcoming problems and to define the requirements of plant breeders.

Author details

Özge Çelik

Address all correspondence to: ocelik@iku.edu.tr

Faculty of Science and Letters, Department of Molecular Biology and Genetics, Istanbul Kultur University, Istanbul, Turkey

References

[1] Jiang G. Molecular markers A2 – Thomas, Brian. In: Murray BG, Murphy DJ, editors. Encyclopedia of Applied Plant Sciences. 2nd ed. Oxford: Academic Press; 2017. pp. 207-214

[2] Tripathy SK, Lenka D, Maharana M, Ithape DM. Biochemical analysis and validation of molecular markers for identification of quality protein maize. Plant Gene. 2018;**14**:69-73

[3] Bosmali I, Ordoudi SA, Tsimidou MZ, Madesis P. Greek PDO saffron authentication studies using species specific molecular markers. Food Research International. 2017;**100**:899-907

[4] Caixeta ET, Ferrão LFV, Maciel-Zambolim E, Zambolim L. Chapter 2 – Molecular Markers, Biotechnology and Plant Breeding. San Diego: Academic Press; 2014. pp. 19-45

[5] Skøt L, Grinberg NF. Genomic selection in crop plants A2 – Thomas, Brian. In: Murray BG, Murphy DJ, editors. Encyclopedia of Applied Plant Sciences. 2nd ed. Oxford: Academic Press; 2017. pp. 88-92

[6] Balázs E. Molecular markers in plant genetics and biotechnology. South African Journal of Botany. 2008;**74**:354

[7] Rocha CML, Vellicce GR, García MG, Pardo EM, Racedo J, Perera MF, de Lucía A, Gilli J, Bogado N, Bonnecarrère V, German S, Marcelino F, Ledesma F, Reznikov S, Ploper LD,

Welin B, Castagnaro AP. Use of AFLP markers to estimate molecular diversity of Phakopsora pachyrhizi. Electronic Journal of Biotechnology. 2015;**18**:439-444

[8] Saha D, Rana RS, Chakraborty S, Datta S, Kumar AA, Chakraborty AK, Karmakar PG. Development of a set of SSR markers for genetic polymorphism detection and interspecific hybrid jute breeding. The Crop Journal. 2017;**5**:416-429

[9] Shabir G, Aslam K, Khan AR, Shahid M, Manzoor H, Noreen S, Khan MA, Baber M, Sabar M, Shah SM, Arif M. Rice molecular markers and genetic mapping: Current status and prospects. Journal of Integrative Agriculture. 2017;**16**:1879-1891

[10] Xu J-Y, Zhu Y, Yi Z, Wu G, Xie G-Y, Qin M-J. Molecular diversity analysis of *Tetradium ruticarpum* (WuZhuYu) in China based on inter-primer binding site (iPBS) markers and inter-simple sequence repeat (ISSR) markers. Chinese Journal of Natural Medicines. 2018; **16**:1-9

[11] Zeinalabedini M, Dezhampour J, Majidian P, Khakzad M, Zanjani BM, Soleimani A, Farsi M. Molecular variability and genetic relationship and structure of Iranian Prunus rootstocks revealed by SSR and AFLP markers. Scientia Horticulturae. 2014;**172**:258-264

[12] Amanullah S, Liu S, Gao P, Zhu Z, Zhu Q, Fan C, Luan F. QTL mapping for melon (*Cucumis melo* L.) fruit traits by assembling and utilization of novel SNPs based CAPS markers. Scientia Horticulturae. 2018;**236**:18-29

[13] Devi EL, Devi CP, Kumar S, Sharma SK, Beemrote A, Chongtham SK, Singh CH, Tania C, Singh TB, Ningombam A, Akoijam R, Singh IM, Singh YR, Monteshori S, Omita Y, Prakash N, Ngachan SV. Marker assisted selection (MAS) towards generating stress tolerant crop plants. Plant Gene. 2017;**11**:205-218

[14] Nayak SN, Singh VK, Varshney RK. Marker-assisted selection A2 – Thomas, Brian. In: Murray BG, Murphy DJ, editors. Encyclopedia of Applied Plant Sciences. 2nd ed. Oxford: Academic Press; 2017. pp. 183-197

[15] Crossa J, Pérez-Rodríguez P, Cuevas J, Montesinos-López O, Jarquín D, de los Campos G, Burgueño J, González-Camacho JM, Pérez-Elizalde S, Beyene Y, Dreisigacker S, Singh R, Zhang X, Gowda M, Roorkiwal M, Rutkoski J, Varshney RK. Genomic selection in plant breeding: Methods, models, and perspectives. Trends in Plant Science. 2017;**22**:961-975

[16] Rashid B, Husnain T, Riazuddin S. Chapter 1 – Genomic approaches and abiotic stress tolerance in plants. In: Ahmad P, Rasool S, editors. Emerging Technologies and Management of Crop Stress Tolerance. San Diego: Academic Press; 2014. pp. 1-37

[17] Li J, Zou X, Zhang L, Cao L, Chen L. Linkage map construction using SSR markers and QTL analyses of stem expansion traits in Brassica juncea. Scientia Horticulturae. 2016;**209**: 67-72

[18] Zhang G, Zhang X, Ye H, Jiang S, Yu H, Li J, Shi Q, Chen G, Zhou Z, Luo J, You X. Construction of high-density genetic linkage maps and QTL mapping in the golden pompano. Aquaculture. 2018;**482**:90-95

[19] Li YH, Reif JC, Hong Hl, Li HH, Liu ZX, Ma YS, Li J, Tian Y, Li YF, Li WB, Qiu LJ. Genome-wide association mapping of QTL underlying seed oil and protein contents of a diverse panel of soybean accessions. Plant Science. 2018;**266**:95-101

[20] Yaobin Q, Peng C, Yichen C, Yue F, Derun H, Tingxu H, Xianjun S, Jiezheng Y. QTL-Seq identified a major QTL for grain length and weight in Rice using near isogenic F2 population. Rice Science. 2018;**25**:121-131

[21] El-Soda M, Malosetti M, Zwaan BJ, Koornneef M, Aarts MGM. Genotype × environment interaction QTL mapping in plants: Lessons from Arabidopsis. Trends in Plant Science. 2014;**19**:390-398

[22] Liu J, Liu B, Cheng F, Liang J, Wang X, Wu J. A high density linkage map facilitates QTL mapping of flowering time in Brassica rapa. Horticultural Plant Journal. 2016;**2**:217-223

[23] Sheng X, Qiu Y, Zhou Y, Zhu W. Joint parameter estimation in the QTL mapping of ordinal traits. Journal of Theoretical Biology. 2017;**432**:100-108

[24] Ganie SH, Upadhyay P, Das S, Prasad Sharma M. Authentication of medicinal plants by DNA markers. Plant Gene. 2015;**4**:83-99

[25] Lavarenne J, Guyomarc'h S, Sallaud C, Gantet P, Lucas M. The spring of systems biology driven breeding. Trends in Plant Science. 2018 (Article in Press)

[26] Poltronieri P. 1 – From Plant Genomics to Plant Biotechnology. Woodhead Publishing; 2013. pp. 3-13

[27] Brotman Y, Riewe D, Lisec J, Meyer RC, Willmitzer L, Altmann T. Identification of enzymatic and regulatory genes of plant metabolism through QTL analysis in Arabidopsis. Journal of Plant Physiology. 2011;**168**:1387-1394

[28] Tohge T, de Souza LP, Fernie AR. Genome-enabled plant metabolomics. Journal of Chromatography B. 2014;**966**:7-20

[29] Langridge P, Fleury D. Making the most of 'omics' for crop breeding. Trends in Biotechnology. 2011;**29**:33-40

[30] Mohanta TK, Bashir T, Hashem A, Abd_Allah EF. Systems biology approach in plant abiotic stresses. Plant Physiology and Biochemistry. 2017;**121**:58-73

[31] Ji L, Neumann DA, Schmitz RJ. Crop epigenomics: Identifying, unlocking, and harnessing cryptic variation in crop genomes. Molecular Plant. 2015;**8**:860-870

[32] Kalavacharla V, Subramani M, Ayyappan V, Dworkin MC, Hayford RK. Chapter 16 – Plant Epigenomics A2 – Tollefsbol, Trygve O, Handbook of Epigenetics. 2nd ed. Academic Press; 2017. pp. 245-258

[33] Tsaftaris A, Kapazoglou A, Darzentas N. 14 – From epigenetics to epigenomics and their implications in plant breeding A2 – Altman, Arie. In: Hasegawa PM, editor. Plant Biotechnology and Agriculture. San Diego: Academic Press; 2012. pp. 207-226

[34] Bhat JA, Ali S, Salgotra RK, Mir ZA, Dutta S, Jadon V, Tyagi A, Mushtaq M, Jain N, Singh PK, Singh GP, Prabhu KV. Genomic selection in the era of next generation sequencing for complex traits in plant breeding. Frontiers in Genetics. 2016;**7**:221

[35] Ray S, Satya P. Next generation sequencing technologies for next generation plant breeding. Frontiers in Plant Science. 2014;**5**:367

[36] Singh VK, Singh AK, Singh S, Singh BD. Next-generation sequencing (NGS) tools and impact in plant breeding. In: Al-Khayri JM, Jain SM, Johnson DV, editors. Advances in Plant Breeding Strategies: Breeding, Biotechnology and Molecular Tools. Cham: Springer International Publishing; 2015. pp. 563-612

[37] Hannon E, Weedon M, Bray N, O'Donovan M, Mill J. Pleiotropic effects of trait-associated genetic variation on DNA methylation: Utility for refining GWAS loci. The American Journal of Human Genetics. 2017;**100**:954-959

[38] Shi X, Ling H-Q. Current advances in genome sequencing of common wheat and its ancestral species. The Crop Journal. 2018;**6**:15-21

[39] Pauwels K, De Keersmaecker SCJ, De Schrijver A, du Jardin P, Roosens NHC, Herman P. Next-generation sequencing as a tool for the molecular characterisation and risk assessment of genetically modified plants: Added value or not? Trends in Food Science & Technology. 2015;**45**:319-326

[40] Fraiture M-A, Herman P, Papazova N, De Loose M, Deforce D, Ruttink T, Roosens NH. An integrated strategy combining DNA walking and NGS to detect GMOs. Food Chemistry. 2017;**232**:351-358

[41] Valdés A, Simó C, Ibáñez C, García-Cañas V. Chapter 13 – Profiling of genetically modified organisms using Omics technologies. In: García-Cañas V, Cifuentes A, Simó C, editors. Comprehensive Analytical Chemistry. Elsevier; 2014. pp. 349-373

[42] Willems S, Fraiture M-A, Deforce D, De Keersmaecker SCJ, De Loose M, Ruttink T, Herman P, Van Nieuwerburgh F, Roosens N. Statistical framework for detection of genetically modified organisms based on next generation sequencing. Food Chemistry. 2016; **192**:788-798

[43] Schmutzer T, Bolger ME, Rudd S, Chen J, Gundlach H, Arend D, Oppermann M, Weise S, Lange M, Spannagl M, Usadel B, Mayer KFX, Scholz U. Bioinformatics in the plant genomic and phenomic domain: The German contribution to resources, services and perspectives. Journal of Biotechnology. 2017;**261**:37-45

Chapter 2

Improving Nitrogen and Phosphorus Efficiency for Optimal Plant Growth and Yield

Lakesh K. Sharma, Ahmed A. Zaeen,
Sukhwinder K. Bali and James D. Dwyer

Additional information is available at the end of the chapter

http://dx.doi.org/10.5772/intechopen.72214

Abstract

Nitrogen (N) and phosphorus (P) are the most important nutrients for crop production. The N contributes to the structural component, generic, and metabolic compounds in a plant cell. N is mainly an essential part of chlorophyll, the compound in the plants that is responsible for photosynthesis process. The plant can get its available nitrogen from the soil by mineralizing organic materials, fixed-N by bacteria, and nitrogen can be released from plant as residue decay. Soil minerals do not release an enough amount of nitrogen to support plant; therefore, fertilizing is necessary for high production. Phosphorous contributes in the complex of the nucleic acid structure of plants. The nucleic acid is essential in protein synthesis regulation; therefore, P is important in cell division and development of new plant tissue. P is one of the 17 essential nutrients for plant growth and related to complex energy transformations in the plant. In the past, growth in production and productivity of crops relied heavily on high-dose application of N and P fertilizers. However, continue adding those chemical fertilizers over time has bad results in diminishing returns regarding no improvement in crop productivity. Applying high doses of chemical fertilizers is a major factor in the climate change in terms of nitrous oxide gas as one of the greenhouse gas and eutrophication that happens because of P pollution in water streams. This chapter speaks about N and P use efficiency and how they are necessary for plant and environment.

Keywords: nitrogen use efficiency, phosphorus, yield, phosphorus and agriculture

1. Introduction

Crop nitrogen use efficiency (NUE) in world cereal production has been estimated to be inefficient with only an average of 33% of fertilized N being recovered during production [1]. Denitrification caused by excessive amount of rainfall and nitrate leaching are the leading causes of N loss in the soil. Loss of N to ground and surface water has resulted from ongoing fertilizer management processes in the Corn Belt region of the USA [2–4]. Insufficient coordination between N applications and the requirement of the crops, applying excessive amounts of N before planting as an example, has been cited as one of the primary reasons for the low NUE of ongoing fertilizer management processes [5, 6]. According to the USDA, for the last two decades, close to 150 kg ha^{-1} has been the usual N application amount in the Corn Belt region of the USA [7], and around 75% of N applications, including the previous fall, was applied before planting [6]. The usual consumption rate of mineral N in the soil for corn for the first 3 weeks after emerging from the ground is less than 0.5 kg ha^{-1} a day. N consumption then increases exponentially to around 3.7 kg ha^{-1} a day after the first 3 weeks until the corn plant reaches the tasselling stage [8]. A recorded consumption of 6 kg ha^{-1} a day has been the highest rate recorded (J.S. Schepers, personal communication). Early season leaching of pre-plant N applications to areas below the crop-rooting zone before the plant reaches its peak N uptake phase is reliant on present soil and weather conditions [9]. The introduction of high amount of available N in the soil profile is risky as it is in danger of being lost to leaching, and the plant can take up denitrification over a period of several weeks before it during its active uptake phase. As the rate of N fertilizer applied in a single pre-plant N application increases, the efficiency of the N application will decrease [10]. However, NUE has been observed to increase when applied in-season as opposed to being applied pre-planting [11]. It has been suggested [12] that N should be applied when required by the crops to increase NUE. Farmer support of the practice of applying N in-season in the corn growing region is low, despite the improved NUE application strategies being supported by ample research [13]. Farmers are likely rejecting the practice of in-season applications in favor of the simpler strategy of applying pre-plant N applications due to the cost and practicality of the labor and equipment associated with in-season applications [6]. Despite the presence of spatial and temporary variables in different landscapes, N is applied in a uniform pattern onto the landscape, ignoring the variables and studies that have proven the economic and environmental benefits of spatially variable N applications, contributing to low NUE in the corn regions [14]. Due to the spatial variabilities in the interior of fields, different sections have varying levels of soil N content, different rates of crop N uptake, and different N responses [15]. Therefore, it is a risk to apply large uniform N pre-plant applications in ignorance of this variability within the field as N in the over-applied areas, or at-risk soils could be lost to environmental factors. Over application of N recommended by out of date N recommendations has been cited as another source of low NUE. Analysis of nitrate in the soil before planting and yield expectations is used as the basis for determining the recommended rate of N application for corn in North Dakota. However, corn will only benefit from these recommendations if they follow a rotation and if manure had been recently used [16]. About 30–60% N loss [17], sometimes N losses could go beyond 70% [18], have been proven by a handful of studies. In North Dakota, the different regional climates, the experience of farmers, and cultural practices are not taken into consideration when developing N recommendations for corn. N availability for corn and the rate of mineralization for residues and organic matter in the soil are dependent on regional climate variables such as the temperature and precipitation. N loss via leaching, the rate of N mineralization, or from denitrification caused by periods of excessive precipitation is affected by different soils in a field with variable traits such as soil texture, pH levels, and organic matter content (OM).

In regard to phosphor the green revolution that followed World War 2, the use of chemical fertilizers increased to increase yields but at the expense of the environment [19]. The common usage of P fertilizers has led to P pollution in the waterways of the USA due to lack of preventative measures to prevent the erosion of P in bodies of water. As a result, wildlife and the environment are at risk. Studies carried out by the Environmental Protection Agency (EPA) has reported and confirmed the presence of P pollution in the Northeastern USA [20]. The application rates of major fertilizers containing N, P, and K have increased in all crops grown in the USA [21]. After 1989, consumption of P temporarily decreased after reports of P erosion in lakes and rivers were released [22] until consumption rates started to increase again in 2010, despite government regulations. Historically, the potato industry in the Northeastern USA was the primary source of P pollution. P fertilizer was applied when not needed, and the potato crops would only recover low amounts of P. It was recently discovered that P concentrations are increasing in lakes and rivers in the Northeast [20], raising concerns about the amount of P currently applied in the agricultural industry in the Northeast. In comparison to potato cultivation in other major potato growing regions, it was found that state-wide P consumption in Maine has declined. This could be attributed to a drop in land dedicated to potato growth over the last 20 years. Despite this, average yield has increased, with the last 2 years of potato production reaching record highs and growing still every year despite declines in P application. However, it was found that this decrease in P application had a nonsignificant reduction. Despite decreasing from 198 to 182 kgha^{-1} (**Figure 1**), it is still very high. When low levels of P are found, the University of Maine Soil Testing Laboratory recommends 50 kgha^{-1}. In the agricultural industry, potato growers apply the maximum amount of fertilizers, making them a prime suspect of being the principal source of P pollution. P pollution in the St. Johns River in Florida has been directly linked to P loss in potato cultivation [23]. The United States Geological Survey found that 71% of the cropland in the USA had at least one of the four contaminations responsible for water quality degradation. Dissolved nitrates, fecal coliform bacteria, suspended sediments, and total P are the four contaminants. A total of 20,000 ha of agricultural land is dedicated to potato cultivation which has a production rate of 44 kg ha^{-1} [21]. The EPA in Maine has raised concerns over the

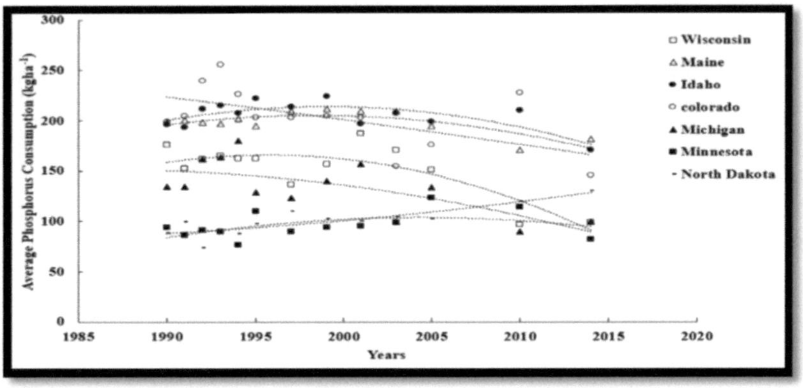

Figure 1. The trend of average P (kg ha^{-1}) used under potato in the key potato-growing states. The polynomial regression analysis was utilized to a potential relationship between years and P use. USDA, National Agricultural Statistics Service, and New England Ag Statistics.

nonpoint source of P that is increasing P pollution in water bodies; 14,407 ha of land has been impaired by P pollution [24]; 3350 t, with an average of 182 kgha^{-1}, of P, was applied to potatoes in 2014. Potatoes have a low P uptake at an average of ~28 kgha^{-1} [25]. Only 10% of P applications are available to potatoes, resulting in a lowered efficiency and loss of P to erosion [26].

In Maine, out of ~3600 t of P that was applied, only 612 t was taken up by potatoes with only 1.12 kg ha^{-1} of it mineralized (fertility and fertilizer book). In Maine, there is P efficiency of ~17%, with applied P only has an efficiency of 16%. P can enter the water via run-in, runoff, or leaching. Water quality degradation is primarily caused by P pollution [66]. Soil runoff and leaching cause an estimated 10–40% of P pollution from agricultural land [46]. Severe eutrophication of water can occur if P concentrations exceed 0.02 ppm [27, 28]. Need for P management to mitigate eutrophication was brought to attention after high levels of P were recorded in the river and lakes of Maine [29]. Despite growers, receiving specific recommendations from soil testing, P pollution is still rising; suggesting growers are still applying excessive P. Because of different parent material, various soil types have different abilities when it comes to releasing available P in soil. Available Ca, Al, and Fe affect the soil ability to hold moisture and the availability of P in soil [30]. There is no clear answer to P requirements, especially in the case of potatoes, despite several studies had been carried out since the 1940s [30–37].

2. Soil and plant analysis

A few studies [38, 39] with the goal of applying the amount of N needed with spatial variables in mind, recommended marking spatial variable management zones (MZ) as part of a soil-based method for variable N applications and for bettering NUE. MZs are defined here as areas within a field with homogeneous characteristics in regard to soil conditions and landscapes. Traits within an MZ such as similar crop yields, electrical conductivity (EC), and producer-defined areas make zones homogenous [40]. Impact of fertilizers on the environment, input-use efficiency, and yield potential are some of the similarities that the attributes have. To define borders for MZ's, a range of methods were put forth by researchers as viable approaches. Geo-referenced data layers (i.e., soil color, electrical conductivity, yield, and topography) are statistically clustered or combined using geospatial statistical analyses within geographic information systems (GIS) to delineate zone boundaries [41]. Soil mapping units [42], remote sensing [41, 43], topography [44], yield maps, and soil EC [45] have been successfully used to delineate the MZ. Static and inconsistent (because of effects of temporal variations on yields) sources is what the MZ relies on for much of the delineation [14]. Because of their static and inconsistent nature, they are likely inappropriate in accounting for all the variability of N requirement within a field.

3. Use of tissue analysis for N management

N concentrations in critical states can be used as an indicator of crop N status. Critical N is the minimum amount of N required to provide the maximum amount of growth at a particular time [46]. The concentration of N is high when the corn plant first starts to grow and develop but eventually decreases as the corn plant matures. Critical N dilution is the graphical depiction of this process [47]. The ratio of actual N in the plant to the critical N set by experiments in the past is called the N nutrition index (NNIN) [58]. The value of NNI more or less than 1 relates to a non-limiting growth or deficient situation of the crop, respectively. Wheat (*Triticum aestivum* L.) [48], grain sorghum (*Sorghum bicolor* L.) [49], rapeseed (*Brassica napus* L.) [50], Rice (*Oryza sativa* L.) [51], and grasses [52] have been used in the NNI approach. Suggested to be the result of competition between corn plants [53] at the early stages of growth, the advent of critical N does not contribute to a solid estimate of crop N status [54]. In what is referred to occasionally as "dilution," an increase in crop biomass will lead to a decrease in N concentrations [53].

4. Spatial variation

Variations in traits such as soils, soil management techniques, production history, movement of water and nutrients, and spatial variation are to classify types of fields for commercial corn production. Because of the spatial variations, changes in N requirements of plants, vulnerability to stress, and productive plant variations across a landscape can occur. Slope changes in the interior of landscape and soil depth and drainage can have huge impacts on grain yield variability and corn grain yield, respectively [55]. Because of the flow of water and deposition of soils containing clay and organic matter into depressions and foot slopes in the landscape in areas of commercial corn production, these landscape features have a high level of N fertility in comparison to the rest of the landscape. The downward shift of these nutrient-rich materials has a noticeable effect on the soil in the upper landscape positions as they have been found to be low in OM [56]. This downward movement also affects P and potassium (K) concentrations as they can be found in higher levels of availability in footholds and depressions. High levels of crop production history naturally lead to higher rates of crop removing, potentially resulting in P and K that are lower than anticipated [44]. This suggests that unlike OM, the redistribution and deposition of soil P and K may not be as strongly related to variations in slope, suggesting a resistance to movement [62]. A loss in growth and yield could be a reaction to crop stress. In some different landscapes, seasonal weather conditions exert an influence on crops. Variations in yield caused by differences in landscape position during dry or wet growing seasons are amplified. High levels of OM or a high water holding capacity can increase the resilience of a landscape to the extreme conditions caused by droughts in comparison to upland areas [56]. Yields can drop if large amount of precipitation causes ponding to occur in depressions in the landscape [56].

5. Fertilizer placement and timing

N application can guarantee the high level of N availability that the crops with high NUE need are required. Injected UAN (urea-ammonium nitrate solutions) has better yield results than the yields that are a result of broadcasting UAN, especially on landscapes with surface residue [58]. Utilizing broadcast UAN applications can result in N loss a variety of ways, including the volatilization of ammonia in the urea portion of the UAN and N immobilization with the surface residue of the landscape [59]. Because of this, the application of fertilizer beneath the surface of the soil may be more efficient. The V7 growth stage in modern corn hybrids accounts for around 15% of total N uptake, as well as 5% of the total dry matter build up [60]; 40% of the total dry matter build up and 60% of the total N uptake have happened by the time the corn plant reaches its silking phase. This means that the period of 30 days between the V7 stage and the VT stage accounts for 40% of the corn plants total N uptake. With no risk of a reduced yield, N synchronization can be enhanced by holding off on in-season applications of N until the V7 stage [61, 62]. At 28 locations with a variety of soils in which timing of N fertilizer application was the experimental variable experimented. At the planting stage, V7 stage, V14 stage, and the silking stage, a single application of ammonium nitrate was put down at a rate of 180 kg N ha^{-1}. At most, of the sites, there was a positive response in corn yield to the N fertilizer. Out of all 28-study sites, only one site experienced slight yield loss when the application of N was held off until the V14 stage. With delayed N applications, there is a possibility that the climate could affect the relative risk of yield loss. Maximum yield was achieved in many locations during dry years by withholding N surface applications until the V14 stage in water-stressed corn. However, the amendment of many of the study sites with animal manure, the use of soybeans as an earlier crop, and the implementation of a variety of tillage systems across the sites have complicated this study. Two locations will be included where corn sites were tilled with the application of manure. The severity and timing of N deficiency due to N mineralization rates and soil N-supplies were affected by the previous crops that were used, manure management, and tillage management. In contradiction to the conclusions in [57, 62], unchangeable yield loss was experienced after N was applied during or after the V6 stage at one of the sites, implicating that at the location, N availability has to be sufficient before side dressing to guarantee that the maximum yield is achieved. There was a decrease in the yield response of the grain to N as N deficiency decreased the longer delay in side-dress N applications, implying that the N deficiency levels were positively interacting with the corn yield at the time of N application.

6. Leaf area index

The ration of the leaf surface area to the ground surface area is called the Leaf Area Index (LAI) [66] and is a direct depiction of the photosynthetic capacity of vegetation [63]. LAI has a direct link to the productivity of vegetation in some species and communities; however, for some, the link between productivity and LAI is dependent on variables such as the canopy extinction coefficient, light, NUE, and the amount of light cut off by the canopy top [64]. C4 plants growing in thick stands having higher NUE and higher leaf area production than C3 plants that are in the same environment is an example of this [64]. Remote sensing has been used to develop approaches for determining LAI. Inversions of canopy radiative transfer models [65] and the empirical relationships between spectral vegetation indices and LAI [66]. A short-coming of algorithms based on vegetation indices is the difficulty in extrapolating their results to larger regions or different canopy types [67]. Vegetation index predictions are often confounded with atmospheric and background effects, canopy architecture, solar-target-sensor geometry, and to lack of spectrum difference when measuring moderate to high levels of LAI [65].

http://dx.doi.org/10.5772/intechopen.72214

7. Environmental interaction

Environmental stress is the primary influence on crop productivity. Corn yields can drop up to and over 70% under negative environmental conditions [68]. Corn hybrids created by breed programs today have shown the ability to withstand environmental stresses, as well as higher plant densities [69]. It is important to note that only 50% of the increases in yield during the modern age of breeding can be attributed to genetic improvements in corn [68], as the other half is a result of better management practices. Corn yield results drop sharply when available soil moisture at depths of 40 cm drops below 25% [70]. Yield can be doubled, however, with the introduction of water via irrigation. During the silking stage, barren ears can occur during drought conditions [71]. Crop yield can drop up to 20% if drought conditions occur after the silking stage [72]. Another study found that moisture stress before silking can cause yield to drop up to 25% and can drop 50% if moisture stress is present during the silking stage [73]. There can be a 21% reduction in yield if soil moisture stress is still present after silking [73]. Moisture stress can cause a plethora of negative symptoms in corn plants such as reduced grain yield, reduced cob length, reduced leaf area, and reduced stem elongation [73]. High temperatures are another source of crop stress. At temperatures of 45° Celsius (113° F), the rate of photosynthesis in corn can be restricted up to 95% during these extreme conditions [74]. Tassel initiation can be postponed by corn stress caused by excessive heat [74]. An increase in high air temperatures to around 32–27°C from a more moderate range of 22–17°C can, respectively, reduce the rate of photosynthesis and the rate of total biomass production by 11 and 32% [75].

8. Spectral response

The spectral properties of leaves can change because environmental stresses [76] observed similar changes in spectral responses across multiple species with changes in plant competition, disease interaction, insufficient ectomycorrhizal infection, senescence, herbicide damage, increased ozone, dehydration, and presence of saline soils. The basis of these responses was that stress reduces chlorophyll content. In regard to the red and green spectrums, chlorophyll α has a low rate of absorbency. Even small changes in chlorophyll concentration can cause increased reflection at these wavelengths [77]. Zhao et al. [78] found more than a 60% reduction in chlorophyll A in leaves after 42 days of emergence, resulting in increased reflectance near 550 and 710 nm. Stress caused by deficiencies in micronutrients is similar to stress caused by N deficiencies. After an evaluation of deficiencies of Fe, S, Mg, and MN, Masoni et al. [79] discovered that decreasing the concentrations of micronutrients caused a decline in chlorophyll concentrations in corn leaves. Chlorophyll a concentrations were 22% less, when Fe, Mg, and Mn were deficient in comparison to unstressed plants. Chlorophyll α concentrations dropped up to 50% when there are deficiencies in sulfur. Because of the decreased concentrations of chlorophyll, there is a decrease in light absorbency, increasing reflectance to around 555 nm and 700 nm [79].

9. Use of spectral properties of plants

The total photosynthetic pigment in a leaf is linked directly to the total amount of solar energy that is absorbed the leaf surface [80]. The photosynthetic potential is directly related to chlorophyll content [81]. Total chlorophyll content changes in response to plant developmental stages or stress. Therefore, measuring chlorophyll content can be a tool for evaluating the physiological health of plants. Gitelson and Merzlyak [82] assessed vegetative indices of a variety of species, leading to a conclusion that the absorption and reflectance of light in the 530–6300 nm and near 700 nm wavelengths were related to chlorophyll content. The light reflectance of plant tissue at the specific wavelengths of 700 and 550 nm was highly correlated with chlorophyll content ($r^2 > 0.97$). Wavelengths in the near infrared spectrum (750–900 nm) were relatively insensitive to chlorophyll content. The ratio of the 750 nm light reflectance to the 550 nm wavelength was used to create an index to be used for predictive measurements [82]. A similar study was conducted on corn [83]. Individual leaves were sampled every 2 weeks. To determine the total chlorophyll content ($r^2 > 0.94$), the red wavelength was used. Crop reflectance is defined as the ratio of the amount of incident light as the denominator to the amount of light reflected back as the numerator [9]. In-season N management was done with active optical sensors by [13] and in the winter wheat fields. During the approach, the NDVI was divided by the growing degree days accrued between planting and sensing. This value was defined as the in-season estimate of yield (INSEY) and was related to the growth rate of the plant. In comparison to solitary sensor readings, INSEY is a better indicator of plant health [13]. To be valid when just using the instrument reading, readings must be done at the same growth stage every year. For developing improved relationships for readings done within a year and over a period of years, time differences between seasons are normalized by INSEY during readings. Blue and red spectra have weaker penetrative properties than green and red-edge spectra when it comes to the capability of light to penetrate into leaves. Eighty percent and higher incident leaf absorption occur in the range of 400–700 nm during the process of photosynthesis [84]. There is a set or range of values, which are not high and narrow in range, in the absorption coefficient in the green and red-edge spectra called saturation, allowing the light in these spectra to be more responsive to changes in the chlorophyll content, especially more than any other wavelength [85]. The ability of leaves of some plant species to absorb light from the visible spectrum increased as plant leaves change their tint from a lighter green to a darker green [80]. The minimum rate of absorption by chlorophyll is 550 nm, while the maximum is 680 nm. Radiation absorption is also influenced by the angle of incident light on the leaf. The comparison of the amount of red light to the amount of near-infrared light absorbed underneath the plant canopy is the most commonly used method of spectral plant analysis [86]. As LAI increases, the amount of light absorbed in the red spectrum and light reflected in the near-infrared [86] increases. LAI could indirectly determine by using a light ratio (675/800) not over but beneath the canopy of the forest. Despite being able to estimate LAI remotely, the authors came to the conclusion that measurement accuracy could be affected by environmental conditions like the angle of incident sunlight and cloud cover. In the evaluation of grass canopies, like approaches have been used [87]. The absorption rate of incident light in spectra (630–690 nm) increases when green biomass increases. Irradiance near the infrared spectrum is defined as lack of absorption or reflection of chlorophyll [87]. Several ratios of the red and near-infrared spectrum are related to the mass of plant greenness [87]. There is a group of ratios that are responsive to physiological parameters and environmental parameters called vegetative indices. Common spectral vegetative indices include chlorophyll indices (Clgreen = $(RNIR/R\ green)^{-1}$) for estimating chlorophyll content [88] and the soil-adjusted vegetation index (SAVI = (RNIRRred) (I + L)/(RNIR + Rred + L)) for LAI estimation [89]. The normalized vegetative index (NDVI) is a widely used vegetative index [13]. Chlorophyll a and b are the most active in the process of photosynthesis, absorbing light (in the red and the blue spectra) and reflecting green spectra [90]. There is more reflectance in the near infrared (700-1400 nm) spectrum of light [90]. Biomass measurements and nutrient deficiencies can be found using these traits in plant leaves [91]. Specialists and researchers prefer to use the Normalized Difference Vegetation Index (NDVI) when they are predicting plant biomasses [91]. NDVI is the ratio of in the red wavelength to NIR light [92]. NDVI = (NIR − red)/(NIR + red), where "NIR" is the reflectance in the near infrared region of the spectrum and "red" is the reflectance in the red region of the spectrum. Because of its usage of the two light spectra and the easiness of its calculations, researchers embrace the NDVI [93].

10. Estimation of vegetative indexes

10.1. Nutrient status

After developing active sensors, the impact of factors such as environmental constraints and ambient light on sampling has been reduced. A plethora of techniques such as destructive plant analysis and soil testing have been used in the past to determine the nutritional status of plants, but recent developments have introduced nondestructive sensors as an alternative [94]. Much of the work done with nondestructive sensors is used to determine the N status of crops [95]. Leaf photosynthesis is negatively impacted and reduced when there is a deficiency of N. Low N availability in corn affects overall production by reducing all components of the corn yield such as kernel dry weight [96]. Crucial for determining the N status of corn, there is a group of wavelengths associated with the N status of corn [97]. Shanahan et al. [98] proposed using NDVI and Green NDVI (GNDVI). In the GDVI, the two spectrums used were NIR, and the other was in the range of 500–600 nm. The light in this spectrum is green; therefore, it was named as green NDVI. The basis for their finding was an experiment of four corn hybrids under irrigation using 5 N rates. Active-optical sensors emitted light in four bands: blue (460 nm), green (555 nm), red (680 nm), and NIR (800 nm). Differences in NDVI were related to N rate and sampling date. N was correlated to increased chlorophyll content ($R^2 > 0.96$). Also, Ref. [99] found that NDVI could be used successfully in evaluating growth and development of small grains.

10.2. Yield estimation

Kitchen and Goulding [104] found it hard to use sensors to establish estimations of yield, even with the established links between green leaf biomass and vegetative indices. In wheat, sensor readings at Feekes growth stage 5 tended to be more correlated with grain yield than any other stage of development [100]. Raun et al. [13] found that sensor-based estimated grain yields were able to explain 83% of grain yield variability. The relationship between sensor reading and yield may be variable over space and time [101]. Inconsistencies have been found in hybrid variations, sampling, seasonal changes, dates, N fertilization, and spatial differences, when determining an estimation of yield [101].

10.3. Nitrogen management using site-specific technologies

Destruction of an area or object is avoided when data are measured via remote sensing methods such as the use of satellite imagery, ground-based active-optical sensors, ground-based reflective sensors, leaf chlorophyll sensors, and aerial imagery or photography [100]. In the agricultural industry, the estimation of land use, land cover, and crop biomass has been done using remote sensing [102]. The in-season status of spatial crop N is now determined using remote sensing [91]. The link between spectral reflectance, crop N status, and chlorophyll content has been better developed as a result of a few studies [91]. Canopy reflectance/color photography, SPAD®(Konica-Minota Americans, Ramsey, NJ), and chlorophyll meters were some of the very first methods of remote sensing used in studies [103]. A plethora of geospatial technologies have been accessible since the mid-1990s for the agricultural market and industry. Crop reflectance, color photography, and GBAO sensors have been successfully used to measure spatial variability in crop canopies.

10.4. Use of sensors and NDVI

When preparing N applications, many farmers use factors such as previous crop, soil management, and soil drainage properties when determining the optimal N rate. However, they commonly do not use in-season tools during these determinations [85]. Farmers apply excessive amounts of N fertilizer in an attempt to guarantee that they will get maximum yield in their fields [105]. Excessive N application leads to problems such as the loss of unused N in the form of nitrate to surface and groundwater, causing environmental problems [105]. Use of proximal plant canopy sensors offers an opportunity for corn producers to adjust N requirement according to the crop requirement. The optimal N rate for any variety of corn and fields is challenging to determine. In order to diminish environmental impact of excess nitrate originating from the production of corn, Schepers et al. [106] suggested that sensing tools to determine to exact amount of N needed instead of applying excessive amounts of N. By estimating crop N status against a standard, the SPAD chlorophyll meter measurement method can help farmers apply N as needed. As a result, farmers still get their maximum yields while using less N fertilizer [107]. However, the SPAD approach requires a laborious process of compiling data from a large number of leaves and then finding a way to standardize N deficient plants from ones that are not deficient with a more significant number of varieties. Active optical sensors are utilized by the SPAD chlorophyll meter to measure two different wavelengths of light (NIR and RED) through the plant leaf. Then, as determined by the manufacturer, a value is computed. The SPAD chlorophyll meter assesses the status of N/nutrition of the plant by analyzing leaf tissue in a nondestructive manner. A positive correlation between chlorophyll content and SPAD chlorophyll meter readings has been proven in multiple studies [108]. However, measurements are done on a one-leaf-at-a-time basis, requiring large of amounts of time to take multiple readings in a field with the SPAD chlorophyll meter. Bullock and Anderson [109] discovered a lack of correlation between V7 stage yields and chlorophyll. An improved correlation between yield and N concentrations in leaves, however, was found at the more advanced stage of R1 and R4. Chlorophyll meter readings at the R1/R4 stages were more closely linked to grain yields than they were to N concentrations in leaves. Correlation coefficients between leaf N and meter readings in the early stages of corn were initially positive ($r^2 = 0.23$), but as the crops grew, there was a drop in value ($r^2 = 0.20$). N recommendations for irrigated corn systems that use irrigation water as a method of N delivery have been successfully made using relative chlorophyll meter readings made by comparing sensor readings from normal farmer fields to readings from plots with high N. Continuous examination of the N status of corn with the chlorophyll meter enabled the additional low N applications when the readings of the chlorophyll meter indicated that N levels had fallen below a set value that determined to be critical [110]. Relative recommendations using the chlorophyll meter require a location where nonlimiting rates of N were applied. Corn grain yield predictions were more accurate when made with relative chlorophyll meter readings rather than predictions using absolute meter readings [105]. Corrective N applications can only be made in a single application in a dryland corn production system, and there are no simple relationships between the application of N that the crop needs and the chlorophyll meter readings [105]. In comparison, low fixed amounts of N can frequently be applied when required in irrigation systems, while guiding N application rates are only active when done with a chlorophyll meter if the meter is the basis for a single N application recommendation [105]. Sripada et al. [113] have analyzed active optical sensors and their possible use at a field scale to determine irrigated corn N status. Variations of growth were manipulated by altering time applications and the rate or amount of N applied. A chlorophyll index (CI) at 590 nm and a NDVI at 590 nm were the two evaluated vegetative indices. Both indices were related to N rate, hybrid, and growth stage. The chlorophyll content during the vegetative growth stages had a stronger relationship to sensor readings than the vegetative reproductive stages. A group of studies has evaluated two available commercial active optical sensors and their efficiency. The two sensors studied were the GS Model 505 (Trimble Inc., Sunnyvale, CA) and the CC ACS-210™ (Holland Scientific, Inc., Lincoln, NE), and they were both used to predict corn yield. The two sensors were differentiated by the wavelengths that they used to calculate NDVI. Both sensors utilized visible and near-infrared wavelengths but the GS Model 505 utilized reflectance measurements from 660 nm and 770 nm, while the CC ACS-210 emitted and detected light at 590 nm and 880 nm. Both sensors are sensitive to crop growth differences ($r^2 > 0.89$). The GS Model 550 exhibited saturation at later stages of growth in comparison to the CC ACS-210, as the different wavelength used by the CC ACS-210 to predict yield reduced its sensitivity and allow usage at the later stages of growth [92]. The GS was also found to be sensitive to the rate of the sensor movement and row spacing [111]. Once again, the CC ACS-210 outperformed the GS by displaying stability during the early and late stages of growth, as well as over multiple row spacing and speed of sensor movement [112]. Therefore, while choosing an appropriate sensor variable N management, the red-edge (680–730 nm) and green wavelength (590 nm) provide a better estimation of canopy development [111]. The hand-held GS 505 is a GBAO sensor, which, unlike the chlorophyll meter, measures reflected light. Satellite imagery, chlorophyll meters, and aerial photography have disadvantages in comparison to the GS when it comes to corn N nutrient management on a field scale regarding speed and labor intensiveness. Ultra-high resolution and fully canopies are needed for aerial photography, while it is not necessary for the GS [113]. Deficiencies of N in plants result in decreased photosynthetic activity, resulting in a higher reflectance of the visible segment of the spectra (400–700), while the stress caused by the N deficiency results in reduced leaf surface area, causing a decrease in NIR (>700 nm) reflectance [114].

10.5. Materials and methods for phosphorus

Three approaches for the study were considered. For the first approach, the last 10-year nutrient analysis data from UMaine Soil Testing Laboratory (UMSTL) were used. Loam, gravelly loam, sandy loam, and silty loam with a parent material of glacial outwash are the soils present in Aroostook County, Maine. Soil testing procedures recommended for the Northeastern USA with publication no.493 by 1:1 method were followed. Modified Morgan soil extracts with inductively coupled plasma (ICP) were used to measure P, Mg, K, Al, and Ca. Using 2874 mL glacial acetic acid mixed with 40 L carboy containing ~ 20 L of distilled water, a modified Morgan extractant (0.62 N NH4OH + 1.25 N CH3COOH) was prepared. Most laboratories did not do bulk density measurements to make it easier for farmers to understand as they convert PPM to pounds/acre. The formula for this is PPM × 2. For all soil testing, the universal assumption/conversion is 2 million pounds or 1000 tons dried and sieved soil per "acre plow layer." Fixed volumes were obtained by scooping rather than weighing by the laboratories to calculate PPM by volume (mg/dm^3) and multiplied by 2 to get a pounds/acre volume. A 1-year N and P study done in 2016 was used for approach 2. A farmer's field in Easton, Aroostook County, was used as the research site for this method. Isotic, frigid Aquic Haplorthods and gravelly loam, fine loamy, isotic, frigid Typic Haplorthods were the soil types used for this study. The Russet Burbank potato cultivar was utilized for this study and was planted 10 cm deep and with row spacing of 91 cm. At planting on the study plots, 6 N treatments, 0, 56, 112, 168, and 280 kgha^{-1}, was done for each of the N fertilizers that are being used in the study, ammonium nitrate (AN) and calcium ammonium nitrate (CAN). The Univ. of Maine Soil Testing Lab., potassium (KCI), gives following recommendations, and P applications were implemented. In the study plot, P was found at a sufficient range (45–49 kgha^{-1}) out of a required range of 24–56 kg ha^{-1} needed which eliminated the need of additional P application. However, the farmer still applied 224 kg ha^{-1} of P on his field leaving the study plot. A UMaine study done in 1996 found no response on the soils, with high P tests (>40 kg ha^{-1}) was cited as the reason for no P applications. The location site was 46 × 46 m and was divided into 3.7 × 9 m subplots. Four replications within a complete randomized block design were used, see **Table 1**.

Two 10 foot potato rows were harvest from each subplot, and each collected bag was graded. The P study from the 1999 master thesis and an article from [115] were reviewed with permission for the third approach (**Table 2**), as well as data from other studies. Maine P study recommendations were developed and critically examined in and near areas in the Northeastern USA and Canada. The Hochmuth study was done in Maine on 12 research locations in farmers in 1995 and 1996.

Location/Soil Sample Depth	OM	pH	P	K	Ca	Mg	N	S	B	Cu	Fe	Mn	Zn
	%	PPM											
Easton/0–15 cm	3.4	5.4	18	386	1065	125	26	133	0.5	1.25	4.9	5.4	1.0
Easton/0–15 cm	3.1	5.5	20	459	1062	114	18	167	0.4	1.19	4.6	6.1	1.0

Soil samples were collected from 0 to 15 cm deep and 15–46 cm deep from the study using a standard soil probe.

Table 1. Before planting at the Easton site, a comprehensive soil test was conducted.

Year	Parameter	pH	OM	P	K	Mg	Ca	Al	B	Cu	Fe	Mn	Na	S	Zn	CEC
			%	ppm												$cmol_c/kg$
2006	Mean	6	4	13	190	98	795	101	0	1	8	6	23	19	1	7
	Median	6	4	13	177	92	714	96	0	1	7	5	23	16	1	7
	Min Range	5	1	1	22	19	132	9	0	0	1	2	3	5	0	2
	Max range	8	9	66	706	262	7596	502	1	20	33	58	81	142	6	17
	Skew	1	1	2	2	1	6	2	1	7	2	4	0	3	2	2
2007	Mean	6	4	14	172	101	909	97	0	1	8	7	24	23	1	7
	Median	6	4	13	161	95	792	92	0	1	7	5	22	17	1	7
	Min Range	5	1	2	34	18	156	16	0	0	2	1	5	1	0	2
	Max range	8	10	34	671	259	21,616	308	1	6	65	152	211	220	25	19
	Skew	0	1	1	2	1	16	1	2	2	3	11	4	3	12	1
2008	Mean	6	4	16	180	101	802	109	0	1	9	7	21	26	1	7
	Median	6	4	16	172	94	731	106	0	1	9	6	20	18	1	7
	Min Range	4	1	3	38	19	116	8	0	0	1	1	7	5	0	3
	Max range	7	7	89	552	228	2770	333	1	11	33	46	82	174	106	15
	Skew	0	0	4	1	1	2	1	2	5	1	3	1	3	30	1
2009	Mean	6	4	17	180	101	896	113	0	1	9	6	27	19	2	7
	Median	6	4	16	170	95	822	110	0	1	8	5	24	15	1	7
	Min Range	4	1	1	28	16	134	19	0	0	2	1	6	4	0	3
	Max range	7	7	95	751	328	4106	301	2	8	37	41	87	240	149	20
	Skew	0	0	5	2	1	2	1	4	3	2	3	1	5	10	1
2010	Mean	6	4	16	182	101	867	108	0	1	9	7	25	20	1	7
	Median	6	4	16	173	94	772	102	0	1	8	6	23	15	1	7
	Min Range	4	1	1	47	10	84	6	0	0	2	2	5	4	0	3
	Max range	8	8	46	592	259	6821	433	2	12	52	49	72	211	29	17
	Skew	0	1	1	1	1	4	2	3	3	3	4	1	4	13	1
2011	Mean	6	4	16	191	111	888	106	0	1	9	6	27	16	1	7
	Median	6	4	16	183	107	795	103	0	1	8	5	25	13	1	7

Year	Parameter	pH	OM	P	K	Mg	Ca	Al	B	Cu	Fe	Mn	Na	S	Zn	CEC
			%	ppm												$cmol_c/kg$
	Min Range	5	1	2	48	23	176	10	0	0	2	1	2	1	0	1
	Max range	7	10	70	687	375	6854	320	2	8	48	37	88	110	20	15
	Skew	0	0	2	1	1	4	1	3	2	2	2	1	3	8	0
2012	Mean	6	4	17	181	113	1004	107	0	1	8	5	25	15	1	7
	Median	6	4	17	175	109	901	101	0	1	7	5	21	12	1	7
	Min Range	5	1	3	42	19	164	17	0	0	1	2	6	4	0	3
	Max range	7	9	80	558	271	4235	288	7	15	33	29	179	244	22	15
	Skew	0	1	2	1	1	2	1	21	5	2	3	2	10	11	0
2013	Mean	6	3	17	172	107	944	113	0	1	9	6	23	17	1	7
	Median	6	3	16	160	105	851	107	0	1	8	5	21	14	1	7
	Min Range	4	1	3	25	17	148	19	0	0	2	1	4	4	0	1
	Max range	8	10	61	488	285	10,358	329	1	6	91	151	77	212	127	13
	Skew	0	1	1	1	1	7	1	1	2	4	15	1	6	26	0
2014	Mean	6	3	17	189	114	989	110	0	1	8	5	20	16	1	7
	Median	6	3	17	178	109	900	106	0	1	8	5	19	14	1	7
	Min Range	5	1	1	31	17	171	20	0	0	2	1	4	4	0	3
	Max range	8	8	65	555	264	4599	358	2	14	43	27	171	225	6	12
	Skew	0	0	2	1	1	3	1	4	4	2	2	4	9	2	0
2015	Mean	6	4	17	198	114	1001	113	0	1	9	5	19	15	1	7
	Median	6	4	17	194	110	897	105	0	1	8	5	17	12	1	7
	Min Range	4	1	2	47	22	176	11	0	0	1	2	2	3	0	2
	Max range	8	10	61	613	392	10,966	515	13	6	53	37	61	90	243	14
	Skew	0	1	2	1	1	7	1	16	2	3	4	1	3	30	1
SR[†]		0.02	0.09	0.46	2.63	2.06	23.97	1.66	0.01	0.03	0.16	0.20	0.87	1.16	0.05	0.07

The number here is the average of ~1000 potato soil samples received by the laboratory each year.

Table 2. The chemical analysis of soil samples received by the UMaine Soil Testing Laboratory each year.

Medium to high P levels was found at all the sites. Diammonium phosphate was applied at 5 P rates, 0, 56, 112, 168, 224 kg P_2O_5 ha^{-1}, using a randomized complete block design with five replications. The "Atlantic" potato cultivar was used for the experiment. All fertilizer was applied at planting. Only one site responded positively to an increase in P rates. To determine the correlation between several parameters of soil that changed with time, the coefficient of correlation (R^2) was used. SAS for Windows 9.2 using PROC REG was used to conduct regression analyses. To compare the N treatments with farmer field yield data and potato yield for approach 2, SAS GLM was used. The relationship between time and P levels was from the UMaine Soil Testing Laboratory who averaged the 10-year data set. The simple percent calculation method was used to calculate the percentage of P samples that were at or above sufficient P levels. The simple percent calculation method is as follows: X = number of samples with P levels above 35 kg/ha and Y = total number of samples.

11. Results and discussion

Of the total Maine soil samples in approach 1, 85% were found to have sufficient P (**Table 2** and **Figure 2**) in the range between 24 and 56 $kgha^{-1}$. However, farmers still applied P in the

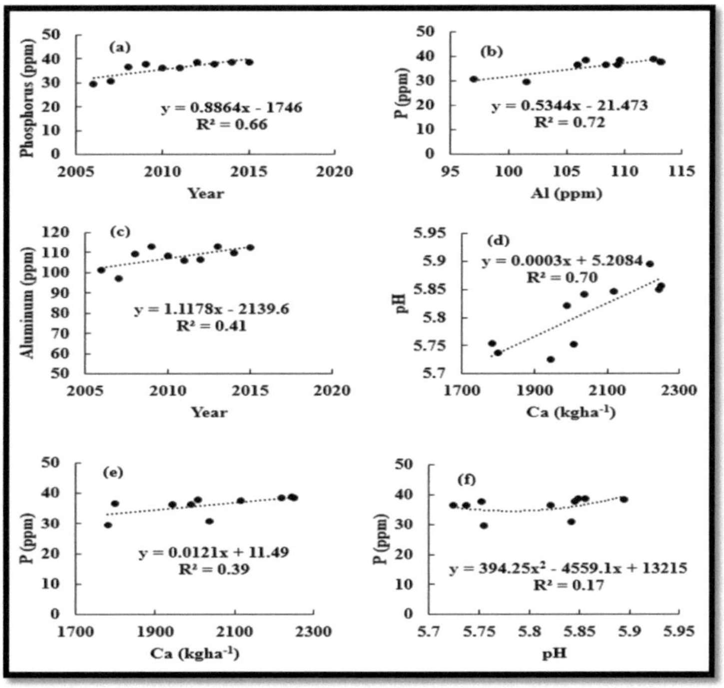

Figure 2. The AL and soil P levels in Aroostook County, Maine. The Univ. of Maine Soil Testing Laboratory has been receiving soil samples since 2006. (a) represents the change in phosphorous levels with time ($p = 0.03$), (b) represents the relationship between Al and P ($p = 0.01$), (c) accounts for the change in AL levels with time ($p = 0.2$), (d) represents the relationship between Ca and pH ($p = 0.02$), (e) represents the relationship between Ca and P($p = 0.2$), and (f) represents the relationship between pH and P ($p = 0.6$). The polynomial model was used in 2 (f) because it was best suited. The trend was positive and properly depicted the significant association of soil P buildup with successive years.

range of 180–200 kg ha^{-1}. Since 2006, growers have been ignorant of recommendations and have been applying significant amounts of P, when the application is not needed, causing P pollution. About 5% of soil samples had more than 56 $kgha^{-1}$ of P, and 10% were P deficient. There may have been a steady build-up of P in the soil over the years ([116, 117]) due to steady P application. In 2016, 85% of the soil samples were found to have a higher range of P in comparison to ~70% in 1996. Growers apply excessive P to protect themselves from P deficiencies caused by soil fixation and erosion in an attempt to ensure that a sufficient supply of P is available to their crops. The low cost of P makes it easier to over apply P. Soil reactive aluminum (Al) that potentially fixes P has a great presence in soil with pH's of around 5–6 pH [118] and is cited by growers as an additional reason to apply excessive P. Maine soil has a general pH range between 4.9 and 6 pH. Al reacts with P to form Al phosphate, a crystalline structure that can transform again to form amorphous Al phosphate [118]. P is also lost to erosion.

The possibility that P might potentially be fixed in high amounts in Maine's soil was confirmed by a gradual increase of Al levels with a coefficient of correlation ($R^2 = 0.41$) over time in the soils of Aroostook County. Despite a strong correlation, the relationship between P, Ca, P, and pH was not significant. However, it was found that P and Al had a very strong relationship with serious correlations ($R^2 = 0.72$). The maximum yield obtained in approach 2 where no P was applied was 59 t ha^{-1} in comparison to the average Maine potato yield of 44 t ha^{-1} [97] with an average P rate of 182 $kgha^{-1}$. Compared with the zero P application at the experimental plot, the farmer applied P at the rate of 224 $kgha^{-1}$ but got a maximum yield of ~53 t ha^{-1}. This confirmed that many farms in Maine potentially have enough P for maximum optimal potato yield. Due to crop and livestock production and high fertilizer applications, soil fertility in Maine may have improved [113]. Another source of improvement in soil fertility is manure application and organic agricultural practices. Over 50% of the annual soil tests in the Northeast States had results that showed high levels of plant-available P [110], indicating that the large P soil reserves could lead to excessive P application as many of the states in the Northeast have not calculated soil P tests results satisfactorily due to P sites that were nonresponsive. Consequently, they were not able to find the optimum P rate for optimal yield. The necessity of developing recommendations for different regions and crop to account for the effects of multiple soil types, climate, crop growth habits, and crop requires has increased the amount of work needed. A study on the effect of residual P in Northeast Florida on the Sebago potato by [107] discovered that soils with P levels greater than 20 mg P kg^{-1} (Mehlich I method) produced about the same yield as soils without P fertilizer. The experiments carried out by [107] were performed on acidic soils with a pH range of 4.5–6, similar to soil pH levels in Maine. Other differences (such as soil types and climate) make it unreasonable to use the results of their study as a basis for P recommendation revisions in Maine. P fertility experiments in the early 1800s in Northern Maine revealed results similar to Rhue's. P applications could potentially be reduced or eliminated without yield reduction on soils that have high amounts of plant-available P (modified Truog method). Potatoes require ~39–45 kg ha^{-1} of P for optimum yields [119, 120]. As potatoes do not use P in soil aggressively, fields with high P concentrations may not need an application of P for several years [121]. However, the variability in P in soil may cause yield to decrease across larger fields. As such, growers may not want to risk nonapplication of P in the soil as it may affect their profits. A study in Florida in 2002 determined that even though P was applied at rates of 0, 12, 24, 49, and 74, yields were not impacted significantly. This may have been due to P fixation in the soil that releases P during plant growth by mineralization or other means [120]. Only one site out of five showed a decrease in yield with a higher P concentration, but it was not significant. The P concentration in the leaves was highly correlated with yield, and only one site found to have an inverse relationship. When graphing all combined outcomes, they are weakly correlated, but individually, they show a strong correlation.

Several studies have indicated that variations in soil type could have an impact on P response regarding crop yield [122, 123] as demonstrated in the introduction. This deems it necessary to study the varying soil types in Maine and Aroostook County, Maine. There are 21 mapping units in Maine, and of these, 15 mapping units are located in Aroostook County, which is a major potato growing area. The soil behaves differently P response of crop yield, P supplying ability, and P retention, and they may vary further in P distribution throughout the landscape. **Table 2** explained that there are soils containing gravel and stones with loam to silt loam. The higher drainage portions of the gravel infused soil may move P into groundwater and nearby streams, whereas silty loam may retain more P. The primary soil order in Maine is Spodosols, susceptible to P deficiency with the third minimum distribution of P among the 12th order after Andisols and Vertisols [124]. A University of Kentucky study on P showed that testing soil P changed under different soils with the same rate of P application [123]. This study explained that different soils have varying rates of P absorption, which results in different levels of P soil tests despite the same rate of application, making it crucial to consider soil type when testing for P levels and recommending P rate for agronomic crops. **Figure 3** explains the rate of change of P concentration in soil depending on the initial test, showing variations in P soil tests with increasing P rates of 16 soil mapping units of large agronomic crops.

Soil pH is a key factor that regulates soil P in soil solutions. The pH range for maximum P availability is between 6 and 7 [112]. At pH levels lower than 6, the available P is fixed by Al and Fe ions and fixed by Ca at a pH higher than 7. Changes in pH were in the study due to its influence. The approach I was used to determine that change in pH over time. Other studies were also discussed to find the answer of the impact of P recommendation and pH on P pollution. Shaver et al. [112] concluded an experiment in Maine to develop P recommendations which proved that there was sufficient P available in Maine soils when only one site ($R^2 = 0.66$) out of 12 was found with positive P response, and the sites with no P response had high to too high P availability. There were not sufficient data to develop P recommendations, so researchers recommend a minimum of ~56 $kgha^{-1}$ when the P value is between 22 and 56 $kgha^{-1}$. Moreover, while the application is not too high, it may have an impact on P erosion to Maine's water sources. Soil pH rates in Maine have improved over the last 10 years (**Figure 4**), mostly after P recommendations were developed. Maine's potato soils have increased after the variety switch from round whites to scab resistant Russet Burbank potatoes and due to grains and other rotation crops that require a higher pH level. The current emphasis on growing grains has led to the increase in soil pH from 5 (20 years ago) to ~6 presently and is expected to continue to improve.

Crop response to an application of P depends on the P availability and crop uptake ability. The soil P can be slowly replenished, but it still depends on the uptake speed and overall crop behavior [125]. Once the crop has absorbed the P from the soil solution, the unavailable or stable form of P can slowly replenish it. The uptake ability also depends on root distribution [125]. P application in potatoes as a banded application that comes in direct contact with the roots ensures P availability later in the season. However, the rainfall before uptake could cause the P to move deeper into the soil or become fixed in unavailable or marginally available forms. In contrast, less movement of P could result in less availability in a banded

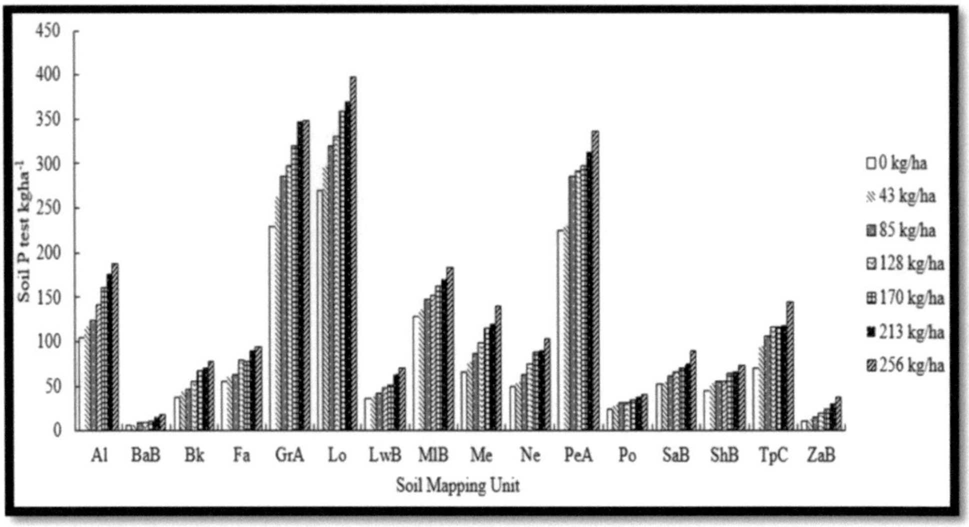

Figure 3. Representing the change in P levels with seven P rates under 16 different soil mapping units in Kentucky, United States. Source: Data adopted from Hochmuth et al. [115].

application as compared to a broadcast application. The potato planting in Maine happens in late May and early June, making the P application more susceptible to erosion due to rainfall. With rainfall in consideration, it is wise to apply P in high doses that are close to first of second hilling (tuber initiation), as the different soil moisture could severely affect the P uptake by the crop plants [122]. Several studies have documented the improvement of crop yields with P application [126]. However, the economic return and response were found only in places with low soil [127]. Inefficiency in soil P application leads to P build up in the soil, particularly when potatoes are used in crop rotation [127]. There is a gap between the rate of P application and the rate of P removal. Potatoes have a relatively high P requirement but a low P uptake behavior [128]. Water can be used as an extracting of P, but due to lack of major leftover undissolved P and analysis difficulties of water as an extracting, several other varieties of extractants have been suggested to extract forms of P in soils used by plants. The Truog Method (1930) is to dilute H_2SO_4 buffered to pH 3.0. The Bray Method is a combination of HCL and NH_4F used to extract acid soluble P forms (mostly Al and Fe bound P [129]) in North Central states. In 1953, another combination (Mehlich 1), HCL and H_2SO_4 acids were introduced to extract P and other nutrients in Southeastern soils. In 1984, Mehlich further expanded on his earlier extractants to Mehlich 3, a combination of acetic acid [HOAc] and nitric acids [HNO_3], salts (ammonium fluoride [NH_4F] and ammonium nitrate [NH_4NO_3]). The standard soil test for P is modified Morgan in Northeastern states due to its acidic soils and low (less than 20) cation exchange capacity (CEC). Modified Morgan used 0.62 M NH4OAc + 1.25 M CH3COOH at pH 4.8. Soil tests show an increase in soil pH in Maine overall with an average P application of ~32 $kgha^{-1}$. The average application of P is between 20 and 50 $kgha^{-1}$, but farmers still apply P to their soils making the excess erode into local water systems. P recommendation studies and their results have never been published making it difficult for growers and researchers alike to amend their practices for better P guidelines.

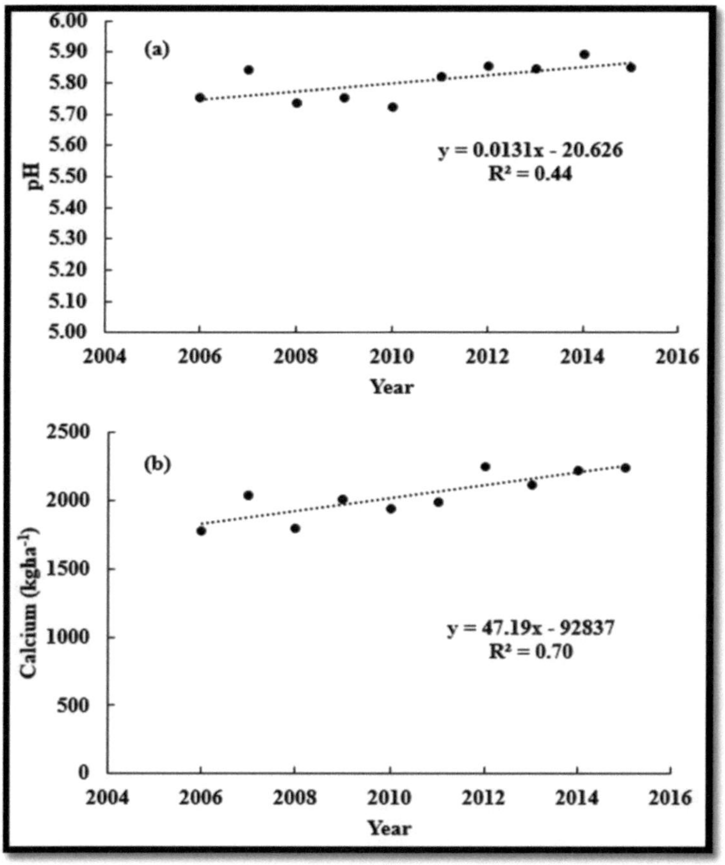

Figure 4. The trend of change in the soil pH and calcium level over time in Aroostook County. (a) The change in pH, and (b) a shift in calcium level with time.

12. Conclusion

Soils in Maine are highly variable and may already possess sufficient P to support the maximum yields of crops. Therefore, the recalibration of the recommendation equation is necessary, by newly inserting low, medium, and high P yield sites. While developing P recommendations, it is important to differentiate between soil types and regions, such as North Dakota State for sunflower, corn, and wheat [130–135]; because soil variability and soil moisture are a driving force toward plant, growth, and nutrient movement among plant roots. The study found that P recommendation needs revision to account for soil variability and a recalibration of the soil P test. The average soil P test has increased showing a buildup of P in Maine soils. Due to unnecessary applications of P, the study recommends a more robust recommendation from low, medium high, and above excellent P level sites. The study also found that types of growers need to be taken into consideration, e.g., table stock growers (not concerned with frying quality), seed producers (no concern of high yield or frying quality), and processing growers (need excellent frying quality with maximum yields) when developing P recommendations. It was also found that an examination needs to be done on the banded application according to the crops root system development as banded applications stay here are applied to the roots that grow beyond the reach of the application.

Author details

Lakesh K. Sharma*, Ahmed A. Zaeen, Sukhwinder K. Bali and James D. Dwyer

*Address all correspondence to: lakesh.sharma@maine.edu

University of Maine Cooperative Extension, Presque Isle, Maine, USA

References

[1] Raun WR, Johnson GV. Improving nitrogen use efficiency for cereal production. Agronomy Journal. 1999;**91**:357-363

[2] Schilling KE. Chemical transport from paired agricultural and restored prairie watersheds. Journal of Environmental Quality. 2002;**31**:1184-1193

[3] Steinheimer TR, Scoggin KD, Kramer LA. Agricultural chemical movement through a field size watershed in Iowa: Surface hydrology and nitrate losses in discharge. Environmental Science & Technology. 1998;**32**:1048-1052

[4] CAST. Gulf of Mexico hypoxia: Land and sea interactions. Task Force Report No. 134. Council for Agricultural Science and Technology, Amaes, IA; 1999

[5] Fageria NK, Baligar VC. Enhancing nitrogen use efficiency in crop plants. Advances in Agronomy. 2005;**88**:97-185

[6] Cassman KG, Dobermann A, Walters DT. Agroecosystems, nitrogen-use efficiency, and nitrogen management. Ambio. 2002;**31**:132-140

[7] United States Department of Agriculture-National Agricultural Statistics Service. Statistics of Fertilizers and Pesticides. Available at http://www.usda.gov/nass/pubs/agr03/03_ch14.pdf (Verified 26 Apr. 2004); 2003

[8] Andrade F, Cirilo AG, Uhart SA, Otegui ME. Ecofisiologia del Cultivo de Maiz. Editorial La Barrosa y Dekalb Press; 1996. p. 292

[9] Schröder JJ, Neeteson JJ, Oenema O, Struik PC. Does the crop or the soil indicate how to save nitrogen in maize production? Reviewing the state of the art. Field Crops Research. 2000;**66**:277-278

[10] Reddy GB, Reddy KR. Fate of nitrogen-15 enriched ammonium nitrate applied to corn. Soil Science Society of America Journal. 1993;**57**:111-115

[11] Olson RA, Raun WR, Chun YS, Skopp J. Nitrogen management and interseeding effects on irrigated corn and sorghum and on soil strength. Agronomy Journal. 1986;**78**:856-862

[12] Keeney DR. Nitrogen Management for Maximum Efficiency and Minimum Pollution. Madison, WI: ASA, CSSA, and SSSA; 1982

[13] Raun WR, Solie JB, Johnson GV, Stone ML, Lukina EV, Thomason WE, Schepers JS. In-season prediction of potential grain yield in winter wheat using canopy reflectance. Agronomy Journal. 2001;**93**:131-138

[14] Lambert DM, Lowenberg-DeBoer J, Malzer GL. Economic analysis of spatial temporal patterns in corn and soybean response to nitrogen and phosphorus. Agronomy Journal. 2006;**98**:43-54

[15] Inman D, Khosla R, Westfall DG, Reich R. Nitrogen uptake across site specific management zones in irrigated corn production systems. Agronomy Journal. 2005;**97**:169-176

[16] Mulvaney RL, Khan SA, Ellsworth TR. Need for a soil-based approach in managing nitrogen fertilizers for profitable corn production. Soil Science Society of America Journal. 2005;**70**:172-118

[17] Bock BR. Efficient use of nitrogen in cropping systems. In: Hauck RD, editor. Nitrogen in Crop Production. Madison, WI: ASA, CSSA, and SSSA; 1984. pp. 273-294

[18] Pierce FJ, Rice CW. Crop rotation and its impact on efficiency of water and nitrogen use. In: Hargrove WL et al., editors. Cropping Strategies for Efficient use of Water and Nitrogen. ASA Special Publ. 51. Madison, WI: ASA, CSSA, and SSSA; 1988. pp. 21-42

[19] IOM (Institute of Medicine) and NRC (National Research Council). A Framework for Assessing Effects of the Food System. Washington, DC: The National Academies Press; 2015

[20] EPA. https://www.epa.gov/aboutepa/epa-region-1-new-england. 2015

[21] USDA-NASS. https: //www.nass.usda.gov/ 2016. [Accessed 1 April 2017]

[22] Bell GL. Lake Erie Chemical and Physical Characteristics Data for 1967. NOAA Data Report ERL GLERL-4, Great Lakes Environmental Research Laboratory, Ann Arbor, MI(PB80-184179); 1980, 9 pp

[23] Livingston-Way P, Hochmuth G, Hanlon E, Tilton A, Bottcher D, Reck B, Konwinski J. Assessment of Agricultural BMPs in the Tri-County Agricultural Area of Northeast Florida. Univ. Fla. Coop. Ext. Serv. Misc. Publ; 1997

[24] Maine Department of Environmental Protection (Maine DEP). Integrated Water Quality Monitoring and Assessment Report. Bureau of Land and Water Quality, Augusta, ME. 2012

[25] Stark JC, Westermann DT, Hopkins BG. Nutrient Management Guidelines for Russet Burbank Potatoes. Univ of Idaho Bull. #840, Moscow, ID. 2004

[26] Van der Zaag P. Soil Fertility Requirements for Potato Production. Technical Information Bulletin 14, CIP, Lima. 1981

[27] Flanagan SM, Nielsen MG, Robinson KW, Coles JF. Water-quality assessment of the New England coastal basins in Maine, Massachusetts, New Hampshire, and Rhode Island: Environmental settings and implications for water quality and aquatic biota. U.S. Geological Survey, Water-Resources Investigations Report 98-4249, Pembroke, New Hampshire. 1999

[28] Sharpley AN, Smith SJ, Jones OR, Berg WA, Coleman GA. The transport of bioavailable phosphorus in agricultural runoff. Journal of Environmental Quality. 1992;**21**:30-35

[29] Sharpley AN. Bioavailable phosphorus in soil. In: Pierzynski GM, editor. Methods for Phosphorus Analysis for Soils, Sediments, Residuals, and Waters. Southern Cooperative Series Bull; 2000. pp. 38-43

[30] McGuire P. Analysis: Belgrade Lake's Water Quality Down. Portland Press Herald. (Posted 27 July 2015); 2015

[31] Peech M. Nutrient status of soils in commercial potato-producing areas of the Atlantic and Gulf Coast: Part II. chemical data on the soils. Soil Science Society of America Journal. 1946;**10**(10):245

[32] Rhue RD, Hensel DR, Yuan TL, Robertson WK. Response of potatoes to soil and fertilizer phosphorus in northeast Florida. The Soil and Crop Science Society of Florida Proceedings. 1981;**40**:58-61

[33] Maier NA, Potocky-Pacay KA, Dahlenburg AP, William CMJ. Effect of phosphorus on the specific gravity of potato tubers (*Solanum tuberosum* L.) of the cultivars Kennebec and Coliban. Australian Journal of Experimental Agriculture. 1989;**29**:869-874

[34] Pierzynski GM, Logan TI. Crop, soil, and management effects on phosphorus soil test levels. Journal of Production Agriculture. 1993;**6**:513-520

[35] Sharpley AN. Dependence of runoff phosphorus on extractable soil phosphorus. Journal of Environmental Quality. 1995;**24**:920-926

[36] Porter GA, McBurnic JC. Crop and soil research. In: Alford AR et al. The Ecology, Economics, and Management of Potato Cropping Systems: A Report of the First Four Years of the Maine Potato Ecosystem Project. Maine Agr. Forest Exp. Sta. Bull. 843. Maine Agr. Forest Exp. Sta. Orono, ME. Maine Agr. Forest Exp. Sta., Orono. ME; 1996. pp. 8-62

[37] Fitzgerald C. Soil phosphorus in Aroostook County (Maine) potato cropping systems: organic matter effects and residual phosphorus contributions. (Master Thesis). Orono: University of Maine; 1998

[38] Franzen DW, Hopkins DH, Sweeney MD, Ulmer MK, Halvorson AD. Evaluation of soil survey scale for zone development of site-specific nitrogen management. Agronomy Journal. 2002;**94**:381-389

[39] Ferguson RB, Lark RM, Slater GP. Approaches to management zone definition for use of nitrification inhibitors. Soil Science Society of America Journal. 2003;**67**:937-947

[40] Kitchen NR, Sudduth KA, Myers DB, Drumond ST, Hong SY. Delineating productivity zones on claypan soil fields using apparent soil electrical conductivity. Computers and Electronics in Agriculture. 2005;**46**:285-308

[41] Schepers AR, Shanahan JF, Liebig MA, Schepers JS, Johnson SH, Jr Luchiari A. Appropriateness of management zones for characterizing spatial variability of soil properties and irrigated corn yields across years. Agronomy Journal. 2004;**96**:195-203

[42] Wibawa WD, Dludlu DL, Swenson LJ, Hopkins DG, Dahnke WC. Variable fertilizer application based on yield goal, soil fertility, and soil map unit. Journal of Production Agriculture. 1993;**6**:255-261

[43] Franzen DW, Wagner G, Sims A. Application of a ground-based sensor to determine N credits from sugarbeet. In Sugarbeet Research and Extension Reports. Vol. 34. Fargo, ND: Sugarbeet Research and Education Board of Minnesota and North Dakota; 2003. pp. 119-123

[44] Kravchenko AN, Bullock DG, Reetz HF. Correlation of corn and soybean grain yield with topography and soil properties. Agronomy Journal. 2000;**92**:75-83

[45] Franzen DW. North Dakota fertilizer recommendation tables and equations. In: NDSU Extension Circular SF-882. Fargo, ND: North Dakota State University Extension Service; 2010

[46] Ulrich A. Physiological bases for assessing the nutritional requirements of plants. Annual Review of Plant Physiology. 1952;**3**:207-228

[47] Greenwood DJ, Lemaire G, Gosse G, Cruz P, Draycott A, Neeteson JJ. Decline in percentage N of C3 and C4 crops with increasing plant mass. Annals of Botany. 1990;**66**:425-436

[48] Justes E, Mary B, Meynard JM, Machet JM, Thelier-Huché L. Determination of a critical nitrogen dilution curve for winter wheat crops. Annals of Botany (London). 1994; **74**:397-407

[49] Van Oosterom EJ, Carberry PS, Muchow RC. Critical and minimum N contents for development and growth of grain sorghum. Field Crops Research. 2001;**70**:55-73

[50] Colnenne C, Meynard JM, Reau R, Justes E, Merrien A. Determination of a critical nitrogen dilution curve for winter oilseed rape. Annals of Botany (London). 1998;**81**:311-317

[51] Sheehy JE, Dionora MJA, Mitchell PL, Peng S, Cassman KG, Lemaire G, Williams RL. Critical nitrogen concentrations: Implications for high-yielding rice(*Oryza sativa* L.) cultivars in the tropics. Field Crops Research. 1998;**59**:31-41

[52] Lemaire G, Salette J. Relation entre dynamique de croissance et dynamique de prélèvement d'azote pour un peuplement de graminées fouragères: I. Etude de l'effect du milieu. Agronomie (Paris). 1984;**4**:423-430

[53] Plénet D, Lemaire G. Relationships between dynamics of nitrogen uptake and dry matter accumulation in maize crops. Plant and Soil. 1999;**216**:65-82

[54] Binford GD, Blackmer AM, Cerrato ME. Relationships between corn yields and soil nitrate in late spring. Agronomy Journal. 1992;**84**:53-59

[55] Kravchenko AN, Robertson GP, Thelen KD, Harwood RR. Management, topographical, and weather effects on spatial variability of crop grain yields. Agronomy Journal. 2005;**97**:514-523

[56] Jiang P, Thelen KD. Effects of soil and topographic properties on crop yield in a north-central corn-soybean cropping system. Agronomy Journal. 2004;**96**:252-258

[57] Ginting D, Moncrieg JF, Gupta SC. Performance of a variable tillage system on interactions with landscape and soil. Precision Agriculture. 2003;**4**:19-34

[58] Fox RH, Kern JM, Piekielek WP. Nitrogen fertilizer source and method and time of application effects on no-till corn yields and N uptakes. Agronomy Journal. 1986;**78**:741-746

[59] Bandel VA, Dzienia S, Stanford G. Comparison of N fertilizers for no-till sites corn. Agronomy Journal. 1980;**72**:337-341

[60] Shanahan JF, Kitchen NR, Raun WR, Schepers JS. Responsive in-season nitrogen management for cereals. Computers and Electronics in Agriculture. 2008;**61**:51-62

[61] Holland KH, Schepers JS. Derivation of a variable rate nitrogen application method for in-season fertilization of corn. Agronomy Journal. 2010;**102**:1415-1424

[62] Scharf PC, Schmidt JP, Kitchen NR, Sudduth KA, Hong SY, Lory JA, Davis JG. Remote sensing for nitrogen management. Journal of Soil and Water Conservation. 2002;**57**:518-524

[63] Whittaker, R. H., and P. L. Marks. 1975. Methods of assessing terrestrial productivity. In: H. Lieth and R. H. Whittaker (Eds.). Primary Productivity of the Biosphere. Ecological Studies 14. Springer-Verlag. pp. 55-118

[64] Anten NPR, Schieving F, Medina E, Werger MJA, Schufflen P. Optimal leaf area indices in C3 and C4 mono- and dicotyledonous species at low and high nitrogen availability. International Journal of Plant Biology. 1995;**95**:541-550

[65] Fang H, Liang S, Kuusk A. Retrieving leaf area index using a genetic algorithm with a canopy radiative transfer model. Remote Sensing of Environment. 2003;**85**:257-270

[66] Chen JM, Cihlar J. Retrieving leaf area index of boreal conifer forests using Landsat TM images. Remote Sensing of Environment. 1995;**55**:153-162

[67] Curran PJ. Multispectral remote sensing for the estimation of green leaf area index. Philosophical Transactions of the Royal Society. 1983;**309**:257-270

[68] Boyer JS. Plant productivity and the environment. Science. 1982;**218**:443-448

[69] Tollenaar M. Genetic improvement in grain yield of commercial maize hybrids grown in Ontario from 1959 to 1988. Crop Science. 1989;**29**:1365-1371

[70] Fulton JM. Relationships among soil moisture stress, plant populations, row spacing and yield of corn. Canadian Journal of Plant Science. 1970;**50**:31-38

[71] Herrero MP, Johnson RR. Drought stress and its effects on corn reproductive systems. Crop Science. 1980;**21**:105-110

[72] Dwyer LM, Stewart DW, Tellenaar M. Analysis of corn leaf photosynthesis under drought stress. Canadian Journal of Plant Science. 1992;**72**:477-481

[73] Denmead OT, Shaw RH. The effects of soil moisture stress at different stages of growth on the development and yield of corn. Agronomy Journal. 1960;**52**:272-274

[74] Crafts-Brandner SJ, Salvucci ME. Sensitivity of photosynthesis in a C4 plant, corn, to heat stress. Plant Physiology. 2002;**129**:1773-1780

[75] Al-Khatib K, Paulsen MG. Photosynthesis and productivity during high temperature stress of wheat genotypes from major world regions. Crop Science. 1990;**30**:1127-1132

[76] Carter GA. Responses of leaf spectral reflectance to plant stress. American Journal of Botany. 1993;**80**:239-243

[77] Carter GA, Knapp AK. Leaf optical properties in higher plants: Linking spectral characteristics to stress and chlorophyll concentration. American Journal of Botany. 2001;**88**:677-684

[78] Zhao D, Reddy KR, Kakani VG, Read JJ, Carter GA. Corn (*Zea mays* L.) growth, leaf pigment concentrations, photosynthesis, and leaf hyperspectral reflectance properties as affected by nitrogen supply. Plant and Soil. 2003;**257**:205-217

[79] Masoni A, Ercoli L, Mariotti M. Spectral properties of leaves deficient in iron, sulfur, magnesium, and manganese. Agronomy Journal. 1996;**88**:937-943

[80] Gates DM, Keegan HJ, Schleter JC, Weidner VR. Spectral properties of plants. Applied Optics. 1964;**4**:11-20

[81] Hatfield JL, Gitelson AA, Schepers JS. Application of spectral remote sensing for agronomic decisions. Agronomy Journal Supplement. 2008;**100**:117-131

[82] Gitelson AA, Merzlyak MN. Remote estimation of chlorophyll content in higher plant leaves. International Journal of Remote Sensing. 1997;**18**:2691-2697

[83] Ciganda V, Gitelson AA, Schepers JS. Non-destructive determination of corn leaf canopy chlorophyll content. Journal of Plant Physiology. 2009;**166**:157-167

[84] Moss RA, Loomis WE. Absorption spectra of leaves. I. The visible spectrum. Plant Physiology. 1952;**27**:370-391

[85] Gitelson AA, Viña A, Arkebauer TJ, Rundquist DC, Keydan G, Leavitt B. Remote estimation of leaf area index and green leaf biomass in maize canopies. Geophysical Research Letters. 2003;**30**:1248

[86] Federer CA, Tanner CB. Spectral distribution of light in the forest. Ecology. 1966;**47**:55-560

[87] Tucker CJ. Red and photographic infrared linear combinations for monitoring vegetation. Remote Sensing of Environment. 1979;**8**:127-150

[88] Gitelson AA, Viña A, Rundquist DC, Ciganda V, Arkebauer TJ. Remote estimation of canopy chlorophyll content in crops. Geophysical Research Letters. 2005;**32**

[89] Huete AR. A soil-adjusted vegetative index (SAVI). Remote Sensing of Environment. 1988;**25**:295-309

[90] Slaton MR, Hunt ER, Smith WK. Estimating near-infrared leaf reflectance from leaf structural characteristics. American Journal of Botany. 2001;**88**:278-284

[91] Osborne SL, Schepers JS, Francis DD, Schlemmer MR. Use of spectral radiance to estimate in-season biomass and grain yield in nitrogen and water stressed corn. Crop Science. 2002;**42**:165-171

[92] Deering DW, Rouse Jr JW, Haas RH, Schell JA. Measuring forage production of grazing units from LANDSAT MSS data. In: Proceedings of the 10th International Symposium on Remote Sensing of Environment; 1975

[93] Deering DW. Rangeland reflectance characteristics measured by aircraft and spacecraft sensors. (Ph.D. dissertation). College Station; 1978

[94] Solari F, Shanahan J, Ferguson RB, Schepers JS, Gitelson AA. Active sensor reflectance measurements of corn nitrogen status and yield potential. Agronomy Journal. 2008; **100**:571-579

[95] Wolfe DW, Henderson DW, Hsiao TC, Alvino A. Interactive water and nitrogen effects on senescence of maize. II. Photosynthetic decline and longevity of individual leaves. Agronomy Journal. 1988;**80**:865-870

[96] Gentry LE, Below FE. Corn productivity as influence by form and availability of nitrogen. Crop Science. 1993;**33**:491-497

[97] Belay A, Claassens AS, Wehner FC. Effect of direct nitrogen and potassium and residual phosphorous fertilizers on soil chemical properties, microbial components and corn yield under long-term crop rotation. Biology and Fertility of Soils. 2002;**35**:420-427

[98] Shanahan JF, Holland K, Schepers JS, Francis DD, Schlemmer MR, Caldwell. Use of crop reflectance sensors to assess corn leaf chlorophyll content; 2003. pp. 129-144. In (ed.)

[99] Hansen PM, Schjoerring JK. Reflectance measurement of canopy biomass and nitrogen status in wheat crops using normalized difference vegetation indices and partial least squares regression. Remote Sensing of Environment. 2003;**86**:542-553

[100] Moges SM, Raun WR, Mullen RW, Freeman KW, Johnson GV, Solie JB. Evaluation of green, red, and near infrared bands for predicting winter wheat biomass, nitrogen uptake, and final grain yield. Journal of Plant Nutrition. 2004;**27**:1431-1441

[101] Inman D, Khosla R, Reich R, Westfall D. Active remote sensing and grain yield in irrigated maize. Precision Agriculture. 2007;**8**:241-252

[102] Henebry GM, de Beurs KM, Gitelson AA. Land surface phenologies of Uzbekistan and Turkmenistan between 1982 and 1999. Arid Ecosystems. 2005;**11**:25-32

[103] Blackmer TM, Schepers JS. Aerial photography to detect nitrogen stress in corn. Journal of Plant Physiology. 1996;**148**:440-444

[104] Kitchen NR, Goulding KW. On-farm technologies and practices to improve nitrogen use efficiency. In: Follett RF, Hatfield JL, editors. Nitrogen in the Environment: Sources, Problems, and Management. Amsterdam, the Netherlands: Elsevier Science; 2001. pp. 335-369

[105] Scharf PC, Brouder SM, Hoeft RG. Chlorophyll meter readings can predict nitrogen need and yield response of corn in the north-central USA. Agronomy Journal. 2006;**98**:655-665

[106] Schepers JS, Varvel GE, Watts DG. Nitrogen and water management strategies to reduce nitrate leaching under irrigated maize. Journal of Contaminant Hydrology. 1995;**20**:227-239

[107] Varvel GE, Schepers JS, Francis DD. Ability for in-season correction of nitrogen deficiency in corn using chlorophyll meters. Soil Science Society of America Journal. 1997; **59**:1233-1239

[108] Schepers JS, Blackmer TM, Francis DD. Predicting N fertilizer needs for corn in humid regions: using chlorophyll meters. In: Bock BR, Kelley KR, editors. Predicting N Fertilizer Needs for Corn in Humid Regions. Bull. Y-226. Muscle Shoals, AL: National Fertilizer and Environmental Research Center; 1992. pp. 105-114

[109] Bullock DG, Anderson DS. Evaluation of the Minolta SPAD-502 chlorophyll meter for nitrogen management in corn. Journal of Plant Nutrition. 1998;**21**:741-755

[110] Shapiro CA. Using a chlorophyll meter to manage nitrogen applications to corn with high nitrate irrigation water. Communications in Soil Science and Plant Analysis. 1999; **30**:1037-1049

[111] Shaver TM, Khosla R, Westfall DG. Evaluation of two Crop Canopy Sensors for Nitrogen Variability Determination in Irrigated Corn. Precision Agriculture. 2011. Online Publication

[112] Shaver TM, Khosla R, Westfal DG. Evaluation of two ground-based active crop canopy sensors in corn: Growth stage, row spacing, and sensor movement speed. Soil Science Society of America Journal. 2010;**74**:2101-2108

[113] Sripada RP, Heiniger RW, White JG, Weisz R. Aerial color infrared photography for determining late-season nitrogen requirements in corn. Agronomy Journal. 2005; **97**:1443-1451

[114] Daughtry CST, Walthall CL, Kim MS, de Colstoun EB, McMurtrey JE. Estimating corn leaf chlorophyll concentration from leaf and canopy reflectance. Remote Sensing of Environment. 2000;**74**:229-239

[115] Bruulsema BTW. Soil fertility in the northeast. The Region. 2006;**90**(1):8-10

[116] Hochmuth G, Weingartner P, Hutchinson C, Tilton A, Jesseman D. Potato yield and tuber quality did not respond to phosphorus fertilization of soils testing high in phosphorus content. HortTechnology. 2002;**12**:420-423

[117] Campbell CA, Zentner RP, Selles F, Jefferson PG, McConkey BG, Lemke R, Blomert BG. Long- term effect of cropping system and nitrogen and phosphorus fertilizer on production and nitrogen economy of grain crops in a Brown Chernozem. Canadian Journal of Plant Science. 2005;**85**:81-93

[118] McKenzie RH, Roberts TL. Soil and fertilizer phosphorus update Alberta Soil Sci. Workshop Proc Coast Terrace Inn Edmonton, AB 84 104 Feb. 20-22; 1990

[119] Pierzynski GM, Sims JT, Vance GF. Soils and Environmental Quality. Chelsea, MI: Lewis Publishers; 1994. p. 313

[120] Locascio SJ, Breland HL. Irish potato yield and leaf composition as affected by dolomite and phosphorus. The Soil and Crop Science Society of Florida Proceedings. 1963;**23**:95-99

[121] Hochmuth G, Maynard D, Vavrina C, Hanlon E. Plant tissue analysis and interpretation for vegetable crops in Florida. Fla. Coop. Ext. Serv. Spec. Ser. SS-VEC 42; 1991

[122] Yuan TL, Hensel DR, Rhue RD, Robertson WK. Nutrient status of a sandy Humaquept under long-term potato cultivation. The Soil and Crop Science Society of Florida Proceedings. 1985;**44**:93-97

[123] Schultz E, Sharma LK, Graham C, Franzen DW. Nitrogen and Phosphorus Recalibration for Modern Varieties of Sunflowers for the Northern Great Plains. Madison, WI. Poster 1121: ASA-CSSA-SSSA; 2015

[124] Cardelli V, Cocco S, Agnelli A, Nardi S, Pizzeghello D, Fernández-Sanjurjo M, Corti G. Chemical and biochemical properties of soils developed from different lithologies in Northwestern Spain (Galicia). Forests. 2017;**8**:135

[125] Thom WO, Dollarhide JE. Phosphorus Soil Test Change Following the Addition of Phosphorus Fertilizer-to-16-Kentucky-Soils. Agronomy-Notes. Paper-11; 2002

[126] Beegle D, Durst PT. Managing phosphorus for crop production. UC055 Penn State Ext. Agrons Facts. 2014;**13**:1-6

[127] Jenkins PD, Ali H. Phosphorus supply and progeny tuber numbers in potato crops. Annals of Applied Biology. 2000;**136**:41-46

[128] Ali H, Khan MA, Shakeel A, Randhawa A. Interactive effect of seed inoculation and phosphorus application on growth and yield of chickpea (*Cicer arietinum* L.). International Journal of Agriculture and Biology. 2004;**6**(1):110-112

[129] Rosen CJ, Kelling KA, Stark JC, Porter GA. Optimizing phosphorus fertilizer management in potato production. American Journal of Potato Research. 2014;**91**(2):145-160

[130] Sharma LK, Franzen DW. Use of corn height to improve the relationship between active optical sensor readings and yield estimates. Precision Agriculture. 2014;**15**:331-345

[131] Franzen DW, Sharma LK, Bu H, Dentond A. Evidence for the ability of active-optical sensors to detect sulfur deficiency in corn. Agronomy Journal. 2016;**108**:2158-2162

[132] Sharma LK, Bu HG, Denton A, Franzen DW. Active-optical sensors using red NDVI compared to red edge NDVI for prediction of corn grain yield in North Dakota, U.S.A. Sensors. 2015;**15**:27832-27853

[133] Bu H, Sharma LK, Denton A, Franzen DW. Sugarbeet root yield and quality prediction at multiple harvest dates using active-optical sensors. Agronomy Journal. 2016; **108**:273-284

[134] Bu H, Sharma LK, Denton A, Franzen DW. Comparison of satellite imagery and ground-based active optical sensors as yield predictors in sugar beet, spring wheat, corn, and sunflower. Agronomy Journal. 2017;**109**:299-308

[135] Sharma LK, Bali SK, Zaeen AA, DwyerJ.D. Potential Reasons of Increased New England States Phosphorus Pollution: A Review. Soil Sci. Soc. A. conf; 2017

Chapter 3

Transgenic Plants: Gene Constructs, Vector and Transformation Method

Lee-Yoon Low, Shun-Kai Yang,
De-Xian Andrew Kok, Janna Ong-Abdullah,
Ngai-Paing Tan and Kok-Song Lai

Additional information is available at the end of the chapter

http://dx.doi.org/10.5772/intechopen.79369

Abstract

The human population has reached 7 billion by 2015 and is estimated to exceed 10 billion by the end of 2050. As such, crops which are the main food source must be produced at a higher pace in order to cater in tandem with the food demand. In the past, traditional plant breeders practice classical breeding techniques to propagate plants with desirable traits. However, traditional breeding technique lies in that only individuals of the same or closely related species can be crossbred. Moreover, traditional breeders will not be able to obtain traits which are not inherent within the gene pool of their target plants through classical breeding. With recent advancements in the field of genetic engineering, it is now possible to insert beneficial genes from a completely different species or even kingdom into a target plant, yielding transgenic plants with multiple ideal traits. To develop a transgenic plant, parameters such as vector constructions, transformation methods, transgene integration, and inheritance of transgene need to be carefully considered to ensure the success of the transformation event. Hence, this chapter aimed to provide an overview of transgenic plants' development, its advantages and disadvantages, as well as its application for the betterment of mankind.

Keywords: genetic engineering, genetic-modified organism (GMO), selectable marker, traits' improvement

1. Introduction

Transgenic plants are plants that have had their genomes modified through genetic engineering techniques either by the addition of a foreign gene or removal of a certain detrimental gene [1]. A foreign gene inserted into a plant can be of a different species or even kingdom. The first transgenic plant was developed through the insertion of *nptII* bacterial antibiotic resistance gene into tobacco [2]. Since then, with the rapid development in plant molecular biology and genetic engineering technology, a wide variety of transgenic plants with important agronomic traits such as pest resistance and drought tolerance have been developed, ranging from dicots to monocots that are amenable to genetic modifications. The main purpose in the production of transgenic plants is to produce crops, which have ideal traits, quality, and high yield. Besides being beneficial to the agriculture sector, the plants are found to be able to act as the factory for pharmaceutical protein production [3].

2. Application of transgenic plants

2.1. Resistance to biotic or abiotic stresses

Biotic stresses occur naturally as a result of stress exerted from other living organism within the same ecosystem. These include bacteria, viruses, herbivores, or native plants [4]. Crop plants are incorporated with disease resistance gene to confer resistance toward these pathogenic diseases that are caused by pest, bacteria, and viruses; this includes tolerance to herbicides. The introduction of genetic modification technology could reduce the usage of expensive pesticides and herbicides in agriculture. The removal of natural pests will lead to a greater yield and better quality of crops. As such, insecticidal toxin genes from a bacterium can be introduced into the plant of interest's genome, thus providing protection to the plant against insect pests [5, 6]. *Bacillus thuringiensis* (*Bt*) crops are an example of transgenic plant produced through this method. In addition, virus-resistant plants can be achieved through the introduction of viral coat proteins into plants [7, 8].

Development of transgenic plants resistant to abiotic stresses is important in this "Global Warming's Terrifying Era". The world climate in the past few decades has changed tremendously culminating in changes to soil composition, humidity, water, sunlight availability, and many other agricultural problems that led to reduction in the crop yield [9, 10]. Hence, genetic engineering technology is needed as a tool to solve these problems by providing the plants with enhanced stress tolerant ability or protection. The manipulation of transcription factors (TFs), late embryogenesis abundant (LEA) proteins, and antioxidant proteins had successfully produced plants tolerant to drought and salinity [11, 12]. Overexpression of the proline biosynthesis enzyme (P5C), which allows the accumulation of osmoprotectant during drought season provides transgenic plants with osmotic stress resistance [13, 14].

2.2. Improving crop yield and nutritional value

Malnutrition is a major health concern that is prevalent especially in the underdeveloped and developing countries due to limited access to nutritious food [15]. Genetic engineering of staple crops has become one of the more effective solutions in addressing this problem. To date, a variety of crops had been successfully modified for better yield as well as for higher nutritional value. Biofortification is a technique used in agriculture to increase the nutritional value of crops. A well-known example would be the golden rice, a variety of *Oryza sativa*, produced to biosynthesize beta-carotene through genetic modification. The golden rice was developed by adding two beta-carotene synthesis genes: phytoene synthase (*psy*) and lycopene β-cyclase (β-*lcy*) (originated from *Narcissus pseudonarcissus*). These genes were driven under the control of the endosperm-specific glutelin promoter together with a bacterial phytoene desaturase (*crt*I, from *Erwinia uredovora*) [16].

2.3. Transgenic plants as bioreactors for recombinant proteins

Plants had been used as a biofactory in the production of the first recombinant human protein in 1989. Product yields from recombinant proteins using mammalian expression systems are low and expensive [17], while bacteria system is incapable of post-translational modification in complex protein formation. Due to this, the production methods had shifted to plant cell systems, which provide cheaper and better alternative sources for recombinant proteins production [18, 19]. The recombinant proteins produced in transgenic plants include antibodies, metabolites or catabolites, proteins, and vaccines [20, 21]. Antibodies and vaccines against gastrointestinal tract diseases, cholera, and malaria are known to be produced in transgenic plants such as potato, banana, algae, and tobacco [22, 23]. An anticancer antibody that recognizes the cells of lung, breast, and colon cancer had also been successfully expressed in rice and wheat seed [24]. However, despite a lot of successful plant-produced antibodies and vaccines, it is difficult to commercialize them and to date, the only plant-produced Newcastle disease vaccine had been approved by the United States Department of Agriculture for poultry farming with several vaccines in clinical trials [25].

3. Gene constructs

A simple functional gene construct consists of a promoter region, gene coding region, and terminator/stop region. In addition, certain gene constructs may contain special sequences such as an enhancer, silencer, or reporter sequences depending on the nature of study. Plant transformation always starts with the transgene construction. Transgene construct generally has similar elements other than the inclusion of the gene of interest and selectable markers. A proper gene construct is crucial for the success of producing ideal transgenic line.

3.1. A typical plant gene

A typical plant gene consists of the regulatory and structural genes [26]. Regulatory genes are usually located at the 5′ upstream of a gene, with its own promoter, enhancer, or silencer region. Structural genes, on the other hand, begin with a catabolite activator protein (cap) site, followed by a leader sequence, start codon, exons, introns, terminator, and a polyadenylation site (poly-A tail). These elements are responsible for DNA transcription. The transcribed pre-mRNA, then undergoes RNA splicing, producing mature mRNA without the introns (non-coding region). This mature mRNA is delivered to the cytoplasm for translation initiated by the binding of ribosomal subunits to the promoter. Translation then begins at the start codon (ATG), with the ribosome moving downstream to the next codon creating a peptide chain with the help of tRNAs and ends once it reaches the chain terminator (stop codon, TAA/TAG).

3.2. Promoters/enhancers

The promoter region is typically located at the 5′ upstream of a gene. Promoters are known for their function in governing gene expression, likened to an on/off switch. In DNA transcription, the promoter sequence is recognized by transcription factors. These transcription factors bind to the consensus region of the promoter and recruit the RNA polymerase. Formation of the RNA polymerase transcription complex marks the beginning of DNA transcription.

The promoters can be categorized into three main groups: constitutive promoters, tissue-specific promoters, and inducible promoters [27]. The constitutive promoters are active at most of the developmental stages, and they directly participate in maintaining moderate and constant level of gene expression. Tissue-specific promoters provide restricted gene expression to certain tissues or gene expression involves in developmental-specific stages. Gene expressions associated with the inducible promoters are greatly affected by environmental stimuli, which allow for the regulation of genes through external factors. **Table 1** shows selected promoters used in plant transformation.

Enhancers are short (50–1500 bp) regions in a gene that can be recognized and bound by activator proteins. These proteins, also referred to as transcription factors, bind to the enhancer, forming an enhancer-bound transcription factor complex, which will later on interact with the mediator complex (TFIID) ultimately aiding in the recruitment of RNA polymerase II. The enhancer-bound transcription factor complex forms a loop and toward the intervening sequence and comes in contact with the promoter region, thus increasing the accessibility of the promoter to the transcription proteins [39]. In contrast, silencers function as the

Promoter	Source	Activity	References
CaMV35S	*Cauliflower mosaic virus*	Constitutive	[28, 29]
Ubiquitin RUBQ1, RUBQ2 and rubi3	Rice	Constitutive	[30]
Ubiquitin Gmubi3	Soybean	Constitutive	[31, 32]
SCR, SRK	*Brassica rapa*	Pollen and stigma specific	[33]
Exo70C2	*Arabidopsis*	Pollen and root specific	[34]
LMW Glu, HMW Glu-1D1	Wheat	Seed specific	[35]
Expansin PcExp2	Sour cherry	Ripened fruits	[36]
Potato class I patatin	Potato	Tuber/storage organ specific	[37]
NtHSP3A	Tobacco	Stress inducible	[38]

Source: Adapted from Hernandez-Garcia et al. [27].

Table 1. Examples of promoter used in plant transformation.

direct opposite of enhancers. Silencers are binding sites for transcription factors known as the repressors. These repressors are known to downregulate the transcription of a gene. In plant genetic engineering, suitable promoter and enhancer are chosen based on the intended regulation of gene expression. Gene expression is kept at basal level when the transgene exerts mild toxicity to the target plant. On the other hand, higher gene expression levels facilitate the detection and monitoring of a transgene which may usually be under expressed in nature.

3.3. Reporter genes

Reporter genes are genes attached to the regulatory sequences or to gene of interest to allow for detection of the transgene expression as well as the localization of expressed proteins [40]. Reporter gene sequences encode proteins or products of the protein after being catalyzed for detection through instruments or simple assays. In contrast, selectable marker genes such as antibiotic genes, herbicidal-resistant genes, and anti-metabolic genes confer resistance toward certain chemical agents, which inhibit nontransgenic plant development [41]. The common reporter genes used to monitor plant transgene expression include green fluorescent protein (GFP), chloramphenicol acetyltransferase (CAT), beta-galactosidase (LacZ), luciferase (Luc), and beta-glucuronidase (GUS). These reporter genes allow differentiation between transformed and nontransformed cells and enable detection of transgene localization and regulation of the expressed and tagged protein. Dual reporter systems such as Luc/Luc and GUS/Luc are also available for better detection in distinguishing proteins [42, 43]. Ideal reporter genes should be highly sensitive, stable, and reliable for large-scale measurements within a wide range of cells and tissues [44]. However, the ideal reporter genes encompassing all the desired properties are still unavailable despite current reporter systems being extensively studied. Each reporter system manifests its own beneficial and detrimental traits. Therefore, due consideration would have to be given when contemplating a suitable reporter gene based on the nature of the study.

3.4. Problem posed by antibiotic resistance reporter genes

Plant transformation techniques available currently are rather efficient but not perfect yet. There are no techniques that are able to provide 100% transformation efficiency. In order to distinguish the transformed and nontransformed plant cells, markers are needed. Antibiotic or herbicide resistance genes act as the primary selective markers in transformant selection to efficiently eliminate the nontransformants [45]. The effectiveness of an antibiotic resistance system is dependent on three criteria: (1) selective agent used should completely inhibit the growth of nontransformed cells, (2) resistance gene is expressed in transformed cells, and (3) explant used for transformation. **Table 2** shows some of the antibiotics used in transgenic plant screening.

Antibiotic screening has provided the initial identification of successful transgenic plant. However, the use of antibiotics always leads to issues on environmental problems and genetic modified (GM) food safety. This is mainly due to the concern of gene pollution when antibiotic gene escapes from the GM plant into the environment through microorganism. Bacteria are known to be able to uptake and integrate foreign DNA pieces into their genome [51]. The microbes surrounding the GM plant might uptake the DNA fragments from the transgenic plant and hence developed resistance to the antibiotics. Besides, gene escape may occur as the

Antibiotics	Mechanism of action	General working concentration (μg/ml)	Selection	References
Kanamycin	Inhibiting ribosomal translocation and eliciting miscoding	50	*nptII*	[46]
Hygromycin B	Inhibit protein synthesis	20–200	*hph*	[47]
Streptomycin	Inhibit protein synthesis	100	*spt*	[48]
Spectinomycin	Inhibit protein synthesis	100	*aadA*	[48]
Phleomycin	DNA breakage	10	*ble*	[49]
Bleomycin	DNA breakage	10	*ble*	[50]

Table 2. Selective antibiotics used for transgenic plants screening.

antibiotic resistance gene can be transferred to the neighboring plants through pollen dispersal [52]. There is also the possibility that consuming the transgenic plant with antibiotic resistance gene may result in the transfer of the genes to the probacteria present in the guts. Hence, the antibiotic resistance marker genes are normally avoided in the transgenic whole plant screening.

Efforts had been made by replacing antibiotic marker gene with another potential marker gene such as reporter genes. Reporter genes such as GFP are not reported as toxic to the environment, instead they are widely used as biosensors. Engineering plants with these reporter genes could prevent the unnecessary buildup of antibiotic resistance in the environment. On the other hand, gene escape can also be avoided through the removal of antibiotic resistance gene from the transgenic plant. The latest genome editing tools such as TALEN and CRISPR/Cas9 may be good tools in the removal of gene markers from the transgenic plants. However, these advanced tools have yet to be shaped.

4. Vectors for the production of transgenic plants

A vector acts as a vehicle that transports the gene of interest into a target cell for replication and expression. Common vector consists of three components: an origin of replication, multicloning site or recombination site, and selectable marker. The origin of replication is an AT-rich region on the vector that initiates the replication of the vector itself by binding to a protein complex, unwinding the vector and thus replicating it with the help of polymerases. The multicloning site is a region that contains multiple unique sequences otherwise known as restriction site that can be cut by specific restriction enzyme, allowing the insertion of the gene of interest. The recombination site allows site-specific recombination to occur between two plasmids. The selectable markers are genetic markers that functions as mentioned in the gene construct section, serving its purpose in validating the insertion of the vector into the *Agrobacterium sp*. In plant transformation, vectors commonly used are Ti plasmid-based vector and plant viral-based vector.

4.1. Plasmid vectors

4.1.1. Ti plasmid

The Ti plasmid is the most commonly used vector in the production of a transgenic plant. The Ti plasmid has an estimated size ranging between 200 and 800 kbp depending on the classes of the Ti plasmid. The Ti plasmid is divided into three main regions: the transfer DNA (T-DNA) region, virulence region, and opine catabolism region. The T-DNA region that is transferred into the plant genome is about 24 kbp in size [53]. This region is bordered by repeat sequences on each end commonly known as the left border and right border. The right border is the critical part essential for the transfer of DNA-causing tumorigenesis. The virulence region, however, is responsible for encoding the *vir* genes, which aids in the transfer of the T-DNA. The T-DNA sequence also codes for opine and phytohormones (auxin and cytokinin) biosynthesis. The three oncogenes (opine, cytokinin, and auxin biosynthesis gene) within the T-DNA are the main causes of tumor formation in plant, leading to the crown gall disease [54]. The growth hormones synthesized are responsible in causing uncontrolled plant cells' proliferation and worsen the situation by enhancing crown gall formation. Opines are the main carbon source utilized by the *A. tumefaciens* that are not naturally synthesized from plant metabolism. Therefore, *A. tumefaciens* will develop its own biosynthetic machinery for production of nutrients by genetically modifying the host cells. The opine catabolism region encodes the genes for proteins involved in opines catabolism. The origin of DNA replication allows stable maintenance of the Ti plasmid in the bacterium. For plant transformation, the Ti plasmid is usually disarmed, with the tumor-inducing genes removed and replaced with the reporter genes together with the gene of interest [55].

The Ti plasmid is large and would become larger with the genes of interest and selectable markers. Large-sized plasmids are cumbersome to handle and have low copy numbers in nature. However, this drawback eventually led to the development of a co-integrative system in combination with the binary vector system which solved the problem for large-sized plasmids.

4.1.2. The co-integrative vector

The co-integrative vector is developed through homologous recombination between an intermediate vector and disarmed Ti plasmid. The intermediate vector is normally the *E. coli* plasmid harboring the gene of interest. Both the intermediate vector and disarmed Ti vector consist of some common sequences, which allow the homologous recombination of the two plasmids to occur. The recombination will result with a large co-integrative vector containing the merged *E. coli* plasmid and disarmed Ti plasmid. This co-integrative vector will later be introduced back into the *Agrobacterium* for transgenic plant transformation. However, the enormous size of the plasmid as a result from the recombination may prove an ominous challenge to be manipulated. Thus, the use of this vector had been discontinued since the binary vector system was introduced.

4.1.3. The binary vector

A two-plasmid system called the binary vector system was developed when researchers found that T-DNA functioned independently without the needs to attach to the Ti plasmid. The binary system involved two plasmids which are the helper vector and mini vector. The mini vector refers to a smaller size plasmid consisting of the T-DNA and the origin of replication of both *E. coli* and *A. tumefaciens*, which allow the plasmid to be cloned in *E. coli* and *A. tumefaciens*. The helper vector refers to a wild-type Ti plasmid without the T-DNA region. The wild-type Ti plasmid is also known as a helper plasmid as it provides the template for all the genes necessary for gene transferring and integration. Both of these helper and mini vectors are introduced together into the *Agrobacterium* and the transformed *Agrobacterium* will be used in plant transformation.

4.2. Plant virus vectors

Viruses are intracellular obligate parasites that require molecular machinery from a specific host to replicate. Viruses have not been found to infect plants through the use of transmission vectors such as aphids, insects, nematodes, and fungi. These viruses have been modified and are used as alternative sources for plant transformation [56]; common plant viruses used in transgenic plant production include the *Cauliflower mosaic virus* (CaMV), *Tobacco mosaic virus* (TMV), *Alfafa mosaic virus* (AMV), *Potato virus X* (PVX), and *Cowpea mosaic virus* (CPMV). The wild-type plant viral vectors have been improved and modified to accommodate their use with *Agrobacteria* as well as the plant host for an increased efficiency level through two approaches. The first approach would be designing virus vectors that are similar to wild types carrying the gene of interest, which are capable of infecting plants.

The second approach would be the development of a 'deconstruct' virus, which occurs through the removal of the undesired viral genes, for example, the coat protein-expressing gene, and to replace them with functional gene such as reporter genes or antibiotic resistance gene, which facilitates transgenic screening.

5. Transformation techniques

Plant transformation refers to the process of altering the genetic constituents in a plant of interest by introducing DNA segments into the plant genome to achieve desired gene expression. Numerous types of plant transformation techniques have now been made accessible to the public. These plant transformation techniques can be categorized under two groups: indirect or direct gene transfer. Indirect gene transfer (also known as vector-mediated gene transfer) involves the introduction of exogenous DNA into the plant genome via biological vectors, whereas direct gene transfer methods involve the introduction of exogenous DNA directly into plant genome through physical or chemical reactions. Different gene transfer methods and their salient features are tabulated in **Table 3**.

Method	Features
Vector-mediated gene transfer	
a. *Agrobacterium*-mediated gene transfer	Efficient to wide range of plants.
b. Plant virus vectors	Efficient and high expression of transgenes.
Direct gene transfer	
a. Physical Methods	
i. Electroporation	Confined to protoplasts that can be regenerated to produce complete and viable plants.
ii. Microinjection	Requires highly skillful technical personnel and limited to one cell per microinjection.
iii. Particle bombardment/microprojectile	Special instrumentation required. High risk of gene rearrangement. May be used for a wide range of plant tissues.
iv. Silicon carbide fibers	Requires careful handling. Requires regenerable cell suspensions.
b. Chemical methods	
v. Polyethylene glycol (PEG)-mediated	Confined to protoplasts. Problems encountered when regenerating these cells into viable plants.
vi. Liposome fusion	Confined to protoplasts which may be regenerated into a viable plant.
vii. Diethylaminoethyl (DEAE) dextran mediated	Does not result in stable transformation.

Table 3. Gene transfer methods in plants and their features.

5.1. *Agrobacterium*-mediated gene transfer

Agrobacterium-mediated transformation is the most common technique used in plant transformation as it is efficient and effective in a wide range of plants. *Agrobacteria* are indigenous to the soil ecosystem. They are pathogenic Gram-negative bacteria that cause crown gall or hairy root disease in plants. The genetic information for tumor growth is encoded on a tumor-inducing plasmid (Ti plasmid) or hairy root-inducing plasmid (Ri plasmid) in the genome of these bacteria. There are generally two types of *Agrobacterium* species that are commonly used in plant transformation; *Agrobacterium tumefaciens* and *Agrobacterium rhizogenes*. *A. tumefaciens* contains the Ti plasmid which causes crown gall disease, whereas *A. rhizogenes* contains the Ri plasmid that causes hairy root disease. The discovery of these two species provides efficient vector systems for the development of transgenic plants when the detrimental genes in *Agrobacteria* are removed. This method had successfully transformed a broad variety of plants such as rice, maize, barley, and tobacco.

A. tumefaciens used for plant transformation are modified *Agrobacteria* which has no tumor-promoting and opine-synthesis genes in their genome. These genes are removed (disarmed) from the bacterial plasmid and replaced with the desired foreign gene or selective markers, making them useful vectors that enables the incorporation of foreign genes into plant's genome, transiently or stably. In order to achieve stable incorporation of genes, the *Agrobacteria* function to transport and integrate the T-DNA into the host's genome through these steps: (1) chemical signal recognition of host, (2) activation of the *vir* gene in *Agrobacterium,* (3) attachment of *Agrobacterium* to plant cells, (4) activation and transportation of virulence proteins, (5) production of T-DNA strand, (6) transfer of T-DNA and virulence protein out of *Agrobacterium,* (7) transfer of T-DNA into plant nuclear, and (8) integration of T-DNA into plant genome. The steps involve in transient transformation are postulated to be identical to the stable transformation with the exception of steps (7) and (8).

Ever since the *Agrobacterium*-mediated transformation protocol had first been introduced, various refinement of the protocol had been ongoing to improve its efficiency. Although traditional *Agrobacterium*-mediated transformation works efficiently in dicotyledonous plants such as potatoes, tomatoes, and tobacco, it is less successful in recalcitrant crops such as wheat and maize due to the lack of wounding response system [57]. This has remained a critical obstacle until the development of plant tissue culture and the introduction of in-planta transformation protocol that has improved the transformation efficiency on these plants at many folds [58].

Agrobacterium-mediated in-planta transformation is a method that does not involve plant tissue culture; transformation is done directly onto a developed plant. This technique includes agroinoculation and agroinfiltration. In agro inoculation, the transformed *Agrobacteria* with the gene of interest is inoculated onto the surface of the plant tissue of a whole plant. It is generally done using either by toothpick, wire loop, or direct organ immersion (floral dip method). Agroinfiltration, on the other hand, can be carried out using syringe or vacuum. Syringe infiltration is simple and cost effective as it injects the transformed *Agrobacteria* onto the underside of the leaf while concurrently ensuring the application of counter pressure on the other side of the leaf. However, it is time consuming and only suitable for small-scale expression.

Vacuum infiltration, however, is rapid and more efficient, thus enabling large-scale production intended commercialization. In this approach, the whole plant is submerged into the transformed *Agrobacteria* suspension with application of a vacuum environment that forces the *Agrobacteria* to penetrate throughout the whole plant. These alternative methods have become popular in plant transformation especially in monocots.

Heat and hydrolysis treatment on target tissues prior to transformation have been reported to enhance transformation efficiency when heat treatment is used, enhancing the efficiency of transformation in different plant tissue such as switchgrass, ryegrass, and rice [59–61]. Similar enhancement had also been obtained in hydrolysis treatment via hydrolytic enzyme such as cellulase, macerase, and pectinase, which provides milder disruption that improved the recovery and regeneration rate of transformed cells [62]. In addition, sonication-assisted plant transformation applied to the target plant tissues prior to *Agrobacteria* immersion or agroinfiltration resulted in effective transformation of recalcitrant plants by creating microwound on explants, which provides better access to *Agrobacterium* [63].

5.2. Direct gene transfers

Direct gene transfer, as the name suggests, involves the direct introduction of exogenous DNA (naked DNA) into the plant nucleus. In order to introduce foreign DNA into the plant cell, the outer membrane of the cell is first disrupted, permeating it for foreign DNA to enter. Most of the methods under direct gene transfer are simple and effective. However, gene expression in these transgenic plants can be transiently or stably transformed.

Direct gene transfer can be categorized into two main groups: physical gene transfer and chemical gene transfer. Physical gene transfer disrupts the cell wall and cell membrane via mechanical means. Among these methods, particle bombardment biolistic is the most common one used in plant transformation since it was first introduced by Sanford et al. [64]. The DNA coated with gold or tungsten particles are shot into the target plant cell under high pressure using a "Gene Gun" (Helios® Bio-Rad). The fast-moving particles allow for the penetration of coated DNA through the thick plant cell wall, directing the foreign DNA into its nucleus. The coated DNA will then separate from the metal particles and integrate itself into the chromosomes within the nucleus of the plant cell. This method had been found to be effective in transforming both dicots and monocots which compensates for the less successful *Agrobacterium*-mediated transformation process. Furthermore, it is also less toxic and applicable to almost all plant cells [65]. The major setbacks of this method, however, lie in the availability of special instruments as well as the delivery efficiency of DNA fragments to the plant nucleus instead of other organelles [66]. In traditional biolistic method, microprojectiles (gold or tungsten) are normally coated with DNA in the presence of calcium chloride and spermidine [67]. The spermidine helps to stabilize the DNA structure and enhances the binding of DNA to the microprojectiles [68]. In the effort of improving this tool, other cationic polyamine such as protamine provides better results when compared with spermidine, as this ensures by protecting the coated DNA from DNase degradation. Biolistic transformation via protamine had been performed in rice and peanut, and the results were shown to be threefolds better when compared to spermidine [69, 70]. Other methods of improving the efficiency of transformation via biolistic guns involved reduction of the amount of DNA coated on the microcarriers [70, 71].

Other physical gene transfer methods include electroporation that uses electrical impulses to facilitate the transfer of foreign DNA into the plant cells. Plant cells are first incubated in a buffer solution containing foreign DNA, followed by the application of electrical impulses into the buffer, resulting in the formation of temporary transient pores on the cell membrane of the plant to allow the foreign DNA to enter. This method is relatively easy and time saving but is only applicable to protoplasts (cell without cell wall). Hence, this method is not commonly practiced in plant transformation.

Chemical gene transfer approaches involves the use of chemical to disrupt cell membrane enabling the entry of foreign DNA. This particular method is not preferable in plant transformation as it is only effective when applied to protoplasts. One of the most prominent chemicals used in this approach is polyethylene glycol (PEG) that is used for destabilizing the cell membrane in the presence of a divalent cation, thus increasing the permeability of the cell membrane, allowing for the uptake of foreign DNA. The exact mechanism for chemical gene transfer is not fully understood, but it was postulated that PEG increases the osmotic pressure and causes contraction in the protoplast; this facilitates endocytosis of the divalent cation/DNA complex [72]. Besides those, liposome is another chemical method that is used in the transformation of plant's protoplast cells. Liposomes act as vehicles to encapsulate and deliver foreign genetic materials into the protoplast. The lipophilic attribute of liposomes provide easy access into the protoplast in transforming the cell [73].

6. Integration and inheritance of the transgenes

6.1. Integration of transgenes

Integration of transgenes into plant cells can be carried out either stably or transiently. In stable transformation, the process normally begins with the introduction of transgenes into the nucleus of plant cells. Stable transformation is achieved when some of the transgenes integrate successfully into the genome of the cell. These transgenes then become a part of the genome and are replicated together, enabling the next generation to inherit and express the transgene. In contrast, transient transformant expressed the transgene transiently, and the transgene is not integrated into the plant genome. In the transiently transformed plant, the copy numbers of transgene inserted remain as they are not replicated. These transgenes are expressed for a limited period of time, and the genes will be lost after several days through cell division. The way how the transgene is expressed in the cell is dependent on the transgene construct design and the method of transformation used. Currently, transient and stable transformation can be achieved through the *Agrobacterium*-mediated method [74] and biolistic method [75]. In the *Agrobacterium*-mediated method, the T-DNA region is inserted into the plant genome forming a stable transformant, whereas the non-integrated T-DNA plasmid expresses the transgene transiently. In the biolistic method or other direct gene transfer methods such as electroporation, transient and stable expression of the transgenes are usually dependent on the plasmid or transgene constructs. Virus-mediated vectors are generally nonintegrative vectors for which transient transformants are frequently produced.

6.2. Inheritance of transgenes

Inheritance of plant genetic information usually obeys the Mendelian law of inheritance in nature. Mendel's first law, the principle of segregation, states that a pair of alleles for each gene will segregate during the formation of gametes, resulting in each gamete harboring only one allele of the gene. Mendel also discovered that the genes of different traits assort independently of each other in the formation of gametes; these genes are passed down to the subsequent progeny generation according to the rules of probability. In addition, the third Mendelian's law states that one allele is dominant to the other allele, which finally determines the corresponding phenotypic attribute of the offspring. However, there are certain cases in which inheritance of a gene does not comply with Mendel's law (non-Mendelian law). These instances include incomplete dominance, codominance, gene controls by multiple alleles, and polytraits.

	Factors
Nature of recipient genome	Genetic background
	Gamete viability
	Chromosome abnormality
	Transformation method
Nature of transgene	Transgene silencing
	Unstable integration of transgene
Interactions between the recipient genome and the transgene	Homozygous lethality
	Poor transmission of transgene
	Mitotic crossover
	Meiotic instability

Source: Adapted from Yin et al. [77].

Table 4. Factors leading to non-Mendelian inheritance of transgene.

Similarly, transgene inheritance may or may not obey the Mendelian law. The rule for transgene inheritance, however, varies due to the location of transgene integration and the copy number of transgenes integrated [76]. Transgene inheritance not obeying the Mendelian law includes deletion of the transgene locus, rearrangement of the inserted transgene, and silencing of the transgene. The factors leading to non-Mendelian inheritance are listed in **Table 4** that had been reviewed by Yin et al. [77]. Overall, the pattern of transgene inheritance is usually analyzed through molecular characterization of the transgene transmission and the segregation analysis of the transgene phenotypic expression pattern.

7. Analysis and confirmation of transgene integration

Analysis and confirmation of transgene integration has to be done through an appropriate method based on the transgene constructs, selectable marker, and reporter gene used. Transgenic plant cells incorporated with antiherbicidal or antibiotic resistance genes are screened by the addition of herbicides or antibiotics to the growing media to distinguish transformed plant cells from the nontransformed plant cells. However, this method requires a large quantity of antibiotics and herbicides that are expensive and worsen by the risk of horizontal gene transfer to other bacteria. Thus, other screening methods such as polymerase chain reaction (PCR) and reporter gene expression screening are used for better accuracy as an alternative screening method for transgenic plants.

Some reporter genes such as the GFP, GUS, and Luc expression are fluorometric or colorimetric, where the expression of these genes could be observed visually or directly under microscopy [78]. Quantifications of the reporter expression are possible with the use of a spectrophotometer. The GUS expression can also be detected through histochemical assay in which the localization of the transgene can be observed. In addition, some of the reporter gene expressions such as CAT and LacZ activity are screened through enzyme assays.

Southern blotting is a molecular method used for the detection of specific DNA sequences within DNA samples. Southern blotting is generally used to identify the number of transgenes inserted into the host genome as well as for the detection of transgene integrity and transgene rearrangement [79, 80]. It is done by cutting the DNA into fragments with endonuclease restriction enzymes, separation by size through electrophoresis, and subsequently transferred onto a nitrocellulose or nylon membrane. Membranes with bound DNA will be incubated in a solution consisting labeled probes, and the pattern of hybridization is detected through autoradiography or via chromogenic detection. The transgene copy number is proportional to the number of bands observed.

The polymerase chain reaction (PCR) method is one of the most sensitive and easiest methods among all the molecular techniques employed for the verification of the transgene. The PCR is generally done with primers specific to the site of plasmid constructs and gene of interest used for development of the transgenic plants. Successful amplification of the DNA fragment with expected band indicates the possible presence of transgene, and this DNA fragment is further confirmed through DNA sequencing. A real-time PCR provides fast, sensitive, and high-throughput molecular PCR–based analysis compared to the traditional Southern blot analysis especially in the area of transgene copy number and zygosity detection in transgenic plants [81]. Real-time PCR is convenient wherein it allows for quantitative, semi-quantitative (qPCR), or qualitative (RT-qPCR) monitoring of target DNA in real time.

In recent years, the emergence of next-generation sequencing (NGS) technologies allows massive parallel generation of sequences from whole genome in a relatively short time with a lower cost. The PCR-based techniques in transgenic analysis often limits by the generation of non-specific products and failure to amplify large exogenous DNA insertion in highly repetitive genomes, multiple insertion, truncated transgene sequences and hinders precise transgene identification [82]. The availability of NGS tools and bioinformatic resources facilitate the study of genome and molecular characterization of complex traits. Besides, the analyses of NGS data allow the identification of precise genomic locations of transgene insertion especially in highly repetitive genome sequence and transposable elements which was not able to be done through the traditional PCR-based method [83]. Hence, NGS approach provides an alternative high-resolution analysis tool for transgenes insertion in GM crops [84].

8. Future directions

GM crops will be a valuable alternative in solving food security problem that happens in a world of growing human population and drastic climate change. However, transgenicity remains a major controversy in the view of biosafety issues spurred by public misconceptions and perceptions to GM plants [85]. In addition, GM crops require years of risk assessments that is time and cost consuming. On the other hand, unintended effects arise could be one of the issues in GM plant production. This is generally due to the transgene integration through illegitimate recombination in plant as the consequences of random transgene integration, gene disruptions, sequence changes, and the production of new proteins [86]. Thus, unintended effects of gene transfer in GM crops should be examined thoroughly through metabolic profiling methods to avoid production of GM plant with significant difference in chemical composition from non-GM plant grown under the same condition.

Recently, the development of engineered site-specific endonucleases such as TFN, TALEN, and CRISPR/Cas9 allows the genetic engineering of plant to be carried out more efficiently and precisely [87]. Problems such as heterozygosity that is commonly faced in agro and gene gun-mediated approaches can be avoided. Hence, the future of transgenic technology is shifting toward the engineered endonuclease genome editing technology. This endonucleases genome editing involves the introduction of a targeted double-stranded DNA breakage (DSB) in genome and consequently stimulating the cellular DNA repair mechanisms. In addition, different genome modification can be done dependent on the types of DSB repair pathways used: (1) non-homologous end joining (NHEJ) and (2) homologous recombination (HR). In NHEJ-mediated genome editing, the target cell self-edits its genome without the addition of foreign gene that may lead to mutation and gene knockout. Since this genome editing is performed without introducing a foreign gene, nontransgenic crops could be obtained. Hence, effort needs to be concerted toward improving the genome editing technology to genetic engineered crops with better agronomic traits and public acceptance.

Acknowledgements

The authors thank Science Fund (02-05-19-SF0026) from the Ministry of Science, Technology and Innovation and Malaysia and Fundamental Research Grant Scheme (FRGS/1/2014/SG05/MOSTI/1) from the Ministry of Higher Education, Malaysia for the support.

Author details

Lee-Yoon Low[1], Shun-Kai Yang[1], De-Xian Andrew Kok[1], Janna Ong-Abdullah[1], Ngai-Paing Tan[2] and Kok-Song Lai[1]*

*Address all correspondence to: laikoksong@upm.edu.my

1 Department of Cell and Molecular Biology, Faculty of Biotechnology and Biomolecular Sciences, Universiti Putra Malaysia, Serdang, Selangor, Malaysia

2 Department of Land Management, Faculty of Agriculture, Universiti Putra Malaysia, Serdang, Selangor, Malaysia

References

[1] Jhansi Rani S, Usha R. Transgenic plants: Types, benefits, public concerns and future. Journal of Pharmacy Research. Aug 2013;**6**(8):879-883

[2] de Framond AJ, Bevan MW, Barton KA, Flavell R, Chilton MD. Mini-Ti plasmid and a chimeric gene construct: New approaches to plant gene vector construction. In: Downey K, Voellmy RW, Ahmad F, editors. Advances in Gene Technology: Molecular Genetics of Plants and Animals. Miami Winter Symp. 20th ed. Florida: Elsevier; 1983. pp. 159-170

[3] Lai KS, Yusoff K, Mahmood M. Heterologous expression of hemagglutinin-neuraminidase protein from Newcastle disease virus strain AF2240 in *Centella asiatica*. Acta Biologica Cracoviensia/Series Botanica. Jan 2012;**54**(1):142-147

[4] Hammond-Kosack K, Jones JD. Responses to plant pathogens. Biochemistry and Molecular Biology of Plants. Jan 2000;**1**(1):1102-1156

[5] Howe GA, Jander G. Plant immunity to insect herbivores. Annual Review of Plant Biology. Jun 2008;**59**(1):41-66

[6] Mithöfer A, Boland W. Plant defense against herbivores: Chemical aspects. Annual Review of Plant Biology. Jun 2012;**63**(1):431-450

[7] Kumar S, Raj SK, Sharma AK, Varma HN. Genetic transformation and development of Cucumber mosaic virus resistant transgenic plants of *Chrysanthemum morifolium* cv. Kundan. Scientia Horticulturae. Feb 2012;**134**(1):40-45

[8] Panter S, Chu PG, Ludlow E, Garrett R, Kalla R, Jahufer MZZ, et al. Molecular breeding of transgenic white clover (*Trifolium repens* L.) with field resistance to Alfalfa mosaic virus through the expression of its coat protein gene. Transgenic Research. Jun 2012;**21**(3):619-632

[9] Datta A. Genetic engineering for improving quality and productivity of crops. Agriculture and Food Security. Nov 2013;**2**(1):15

[10] Atkinson NJ, Urwin PE. The interaction of plant biotic and abiotic stresses: From genes to the field. Journal of Experimental Botany. Jun 2012;**63**(10):3523-3543

[11] Umezawa T, Fujita M, Fujita Y, Yamaguchi-Shinozaki K, Shinozaki K. Engineering drought tolerance in plants: Discovering and tailoring genes to unlock the future. Current Opinion in Biotechnology. Apr 2006;**17**(2):113-122

[12] Bhatnagar-Mathur P, Vadez V, Sharma KK. Transgenic approaches for abiotic stress tolerance in plants: Retrospect and prospects. Plant Cell Reports. Mar 2008;**27**(3):411-424

[13] Bray EA. Responses to abiotic stresses. In: Gruissem W, Buchannan B, Jones R, editors. Biochemistry and Molecular Biology of Plants. Maryland: American Society of Plant Physiologists; 2000. pp. 1158-1203

[14] Chaves MM, Maroco JP, Pereira JS. Understanding plant responses to drought—From genes to the whole plant. Functional Plant Biology. Mar 2003;**30**(3):239-264

[15] Pérez-Massot E, Banakar R, Gómez-Galera S, Zorrilla-López U, Sanahuja G, Arjó G, et al. The contribution of transgenic plants to better health through improved nutrition: Opportunities and constraints. Genes & Nutrition. Jan 2013;**8**(1):29-41

[16] Beyer P, Al-Babili S, Ye X, Lucca P, Schaub P, Welsch R, et al. Golden Rice: Introducing the beta-carotene biosynthesis pathway into rice endosperm by genetic engineering to defeat vitamin A deficiency. The Journal of Nutrition. Mar 2002;**132**(3):506S-510S

[17] Ganz PR, Dudani AK, Tackaberry ES, Sardana R, Sauder C, XiongYing C, et al. Expression of human blood proteins in transgenic plants: The cytokine GM-CSF as a model protein. In: Owen MRL, Pen J, editors. Transgenic Plants: A Production System for Industrial and Pharmaceutical Proteins. New Jersey: Wiley; 1996. pp. 281-297

[18] Fischer R, Emans N. Molecular farming of pharmaceutical proteins. Transgenic Research. Aug 2000;**9**(4-5):279-299

[19] Goldstein DA, Thomas JA. Biopharmaceuticals derived from genetically modified plants. QJM: An International Journal of Medicine. Nov 2004;**97**(11):705-716

[20] Doran PM. Therapeutically important proteins from in vitro plant tissue culture systems. Current Medicinal Chemistry. Mar 2013;**20**(8):1047-1055

[21] Fischer R, Schillberg S, Buyel JF, Twyman RM. Commercial aspects of pharmaceutical protein production in plants. Current Pharmaceutical Design. Sep 2013;**19**(31):5471-5477

[22] Gregory JA, Topol AB, Doerner DZ, Mayfield S. Alga-produced cholera toxin-Pfs25 fusion proteins as oral vaccines. Applied and Environmental Microbiology. Jul 2013; **79**(13):3917-3925

[23] Wang L, Webster DE, Campbell AE, Dry IB, Wesselingh SL, Coppel RL. Immunogenicity of Plasmodium yoelii merozoite surface protein 4/5 produced in transgenic plants. International Journal for Parasitology. Jan 2008;**38**(1):103-110

[24] Stöger E, Vaquero C, Torres E, Sack M, Nicholson L, Drossard J, et al. Cereal crops as viable production and storage systems for pharmaceutical scFv antibodies. Plant Molecular Biology. Mar 2000;**42**(4):583-590

[25] Takeyama N, Kiyono H, Yuki Y. Plant-based vaccines for animals and humans: Recent advances in technology and clinical trials. Therapeutic Advances in Vaccines. Sep 2015; **3**(5-6):139-154

[26] Debnath M, Prasad GBKS, Bisen PS. Molecular Diagnostics: Promises and Possibilities. Dordrecht: Springer Netherlands; 2010

[27] Hernandez-Garcia CM, Finer JJ. Identification and validation of promoters and cis-acting regulatory elements. Plant Science. Mar 2014;**217**(1):109-119

[28] Odell JT, Nagy F, Chua NH. Identification of DNA sequences required for activity of the cauliflower mosaic virus 35S promoter. Nature. Feb 1985;**313**(6005):810-812

[29] Lim YY, Lai KS. Generation of transgenic rice expressing cyclotide precursor Oldenlandia affinis kalata B1 proetin. Journal of Applied Pharmaceutical Science. 2017;**27**(2):667-671

[30] Sivamani E, Qu R. Expression enhancement of a rice polyubiquitin gene promoter. Plant Molecular Biology. Jan 2006;**60**(2):225-239

[31] Chiera JM, Bouchard RA, Dorsey SL, Park E, Buenrostro-Nava MT, Ling PP, et al. Isolation of two highly active soybean (*Glycine max* (L.) Merr.) promoters and their characterization using a new automated image collection and analysis system. Plant Cell Reports. Aug 2007;**26**(9):1501-1509

[32] Hernandez-Garcia CM, Martinelli AP, Bouchard RA, Finer JJ. A soybean (Glycine max) polyubiquitin promoter gives strong constitutive expression in transgenic soybean. Plant Cell Reports. May 2009;**28**(5):837-849

[33] Lai KS, Masatsugu T. Isolation and characterization of an Arabidopsis thaliana self-incompatibility mutant induced by heavy-ion beam irradiation. Acta Biologica Cracoviensia/Series Botanica. Jan 2013;**55**(2):146-152

[34] Lai KS. Analysis of EXO70C2 expression revealed its specific association with late stages of pollen development. Plant Cell, Tissue and Organ Culture. Jan 2016;**124**(1):209-215

[35] Kawakatsu T, Takaiwa F. Cereal seed storage protein synthesis: Fundamental processes for recombinant protein production in cereal grains. Plant Biotechnology Journal. Dec 2010;**8**(9):939-953

[36] Karaaslan M, Hrazdina G. Characterization of an expansin gene and its ripening-specific promoter fragments from sour cherry (*Prunus cerasus* L.) cultivars. Acta Physiologiae Plantarum. Nov 2010;**32**(6):1073-1084

[37] Ihemere U, Arias-Garzon D, Lawrence S, Sayre R. Genetic modification of cassava for enhanced starch production. Plant Biotechnology Journal. Jul 2006;**4**(4):453-465

[38] Navarre C, Sallets A, Gauthy E, Maîtrejean M, Magy B, Nader J, et al. Isolation of heat shock-induced Nicotiana tabacum transcription promoters and their potential as a tool for plant research and biotechnology. Transgenic Research. Aug 2011;**20**(4):799-810

[39] Ong C-T, Corces VG. Enhancer function: New insights into the regulation of tissue-specific gene expression. Nature Reviews. Genetics. Apr 2011;**12**(4):283-293

[40] Seymour PA, Sander M. Immunohistochemical detection of beta-galactosidase or green fluorescent protein on tissue sections. Methods in Molecular Biology. Jan 2007;**411**(1): 13-23

[41] Ziemienowicz A. Plant selectable markers and reporter genes. Acta Physiologiae Plantarum. Sep 2001;**23**(3):363-374

[42] Koo J, Kim Y, Kim J, Yeom M, Lee IC, Nam HG. A GUS/luciferase fusion reporter for plant gene trapping and for assay of promoter activity with luciferin-dependent control of the reporter protein stability. Plant & Cell Physiology. Aug 2007;**48**(8):1121-1131

[43] Matsuo N, Minami M, Maeda T, Hiratsuka K. Dual luciferase assay for monitoring transient gene expression in higher plants. Plant Biotechnology. Jan 2001;**18**(1):71-75

[44] Santi C, Svistoonoff S, Constans L, Auguy F, Duhoux E, Bogusz D, et al. Choosing a reporter for gene expression studies in transgenic actinorhizal plants of the Casuarinaceae family. Plant and Soil. Jul 2003;**254**(1):229-237

[45] Wong CKF, Lai KS, Yun WM, Mahmood M. Minimal inhibitory concentration of common selectable agents on recalcitrant malaysian rice cultivar. Asian Journal of Agricultural Research. Feb 2016;**10**(2):126-131

[46] Pantazis CJ, Fisk S, Mills K, Flinn BS, Shulaev V, Veilleux RE, et al. Development of an efficient transformation method by *Agrobacterium tumefaciens* and high throughput spray assay to identify transgenic plants for woodland strawberry (Fragaria vesca) using NPTII selection. Plant Cell Reports. Mar 2013;**32**(3):329-337

[47] Boyer L, Roussel X, Courseaux A, Ndjindji OM, Lancelon-Pin C, Putaux J-L, et al. Expression of *Escherichia coli* glycogen branching enzyme in an Arabidopsis mutant devoid of endogenous starch branching enzymes induces the synthesis of starch-like polyglucans. Plant, Cell & Environment. Jul 2016;**39**(7):1432-1447

[48] Bansal KC, Singh AK. Plastid transformation in eggplant. In: Maliga P, editor. Chloroplast Biotechnology. New Jersey: Humana Press; 2014. pp. 305-316

[49] Ma X, Pan K, Zhang L, Zhu B, Yang G, Zhang X. Genetic transformation of *Nannochloropsis oculata* with a bacterial phleomycin resistance gene as dominant selective marker. Journal of Ocean University of China. Apr 2016;**15**(2):351-356

[50] Hille J, Verheggen F, Roelvink P, Franssen H, van Kammen A, Zabel P. Bleomycin resistance: A new dominant selectable marker for plant cell transformation. Plant Molecular Biology. May 1986;**7**(3):171-176

[51] de la Cruz F, Davies J. Horizontal gene transfer and the origin of species: Lessons from bacteria. Trends in Microbiology. Mar 2000;**8**(3):128-133

[52] Morris WF, Kareiva PM, Raymer PL. Do barren zones and pollen traps reduce gene escape from transgenic crops? Ecological Applications. Feb 1994;**4**(1):157-165

[53] Barker RF, Idler KB, Thompson DV, Kemp JD. Nucleotide sequence of the T-DNA region from the *Agrobacterium tumefaciens* octopine Ti plasmid pTi15955. Plant Molecular Biology. Nov 1983;**2**(6):335-350

[54] Christie PJ, Gordon JE. The Agrobacterium Ti plasmids. Microbiology Spectrum. Dec 2014;**2**(6):295-313

[55] Van Montagu M, Zambryski P. Agrobacterium and Ti plasmids. In: Brenner's Encyclopedia of Genetics. Amsterdam: Elsevier; 2013. pp. 55-57

[56] Casteel CL, Falk BW. Plant virus-vector interactions: More than just for virus transmission. In: Current Research Topics in Plant Virology. Cham: Springer International Publishing; 2016. pp. 217-240

[57] Potrykus I. Gene transfer to plants: Assessment of published approaches and results. Annual Review of Plant Physiology and Plant Molecular Biology. Jun 1991;**42**(1):205-225

[58] Jan SA, Shinwari ZK, Shah SH, Shahzad A, Zia MA, Ahmad N. In-planta transformation: Recent advances. Romanian Biotechnology Letters. Jan 2016;**21**(1):11085-11091

[59] Chen X, Equi R, Baxter H, Berk K, Han J, Agarwal S, et al. A high-throughput transient gene expression system for switchgrass (*Panicum virgatum* L.) seedlings. Biotechnology for Biofuels. May 2010;**3**(1):9

[60] Patel N, Patel P, Kumari U, Kendurkar SV, Khan BM. Microprojectile bombardment assisted agroinfection increases transformation efficiency of *Withania somnifera* (L.). Research in Biotechnology. 2014;**5**:13-24

[61] Wong CKF, Lai KS, Wong MY, Maziah M. Heat and hydrolytic enzymes treatment improved the Agrobacterium-mediated transformation of recalcitrant indica rice (*Oryza sativa* L.). Plant Cell, Tissue and Organ Culture. Apr 2016;**125**(1):183-190

[62] Weber S, Friedt W, Landes N, Molinier J, Himber C, Rousselin P, et al. Improved Agrobacterium-mediated transformation of sunflower (*Helianthus annuus* L.): Assessment of macerating enzymes and sonication. Plant Cell Reports. Jan 4, 2003;**21**(5):475-482

[63] Dutta I, Kottackal M, Tumimbang E, Tajima H, Zaid A, Blumwald E. Sonication-assisted efficient Agrobacterium-mediated genetic transformation of the multipurpose woody desert shrub *Leptadenia pyrotechnica*. Plant Cell, Tissue and Organ Culture. Mar 2013; **112**(3):289-301

[64] Sanford JC. Biolistic plant transformation. Physiologia Plantarum. May 1990;**79**(1):206-209

[65] Lai KS, Abdullah P, Yusoff K, Mahmood M. An efficient protocol for particle bombardment-mediated transformation of *Centella asiatica* callus. Acta Physiologiae Plantarum. Nov 2011;**33**(6):2547-2552

[66] Furth PA. Gene transfer by biolistic process. Molecular Biotechnology. Apr 1997; **7**(2):139-143

[67] Klein TM, Kornstein L, Sanford JC, Fromm ME. Genetic transformation of maize cells by particle bombardment. Plant Physiology. Sep 1989;**91**(1):440-444

[68] Thomas TJ, Kulkarni GD, Greenfield NJ, Shirahata A, Thomas T. Structural specificity effects of trivalent polyamine analogues on the stabilization and conformational plasticity of triplex DNA. The Biochemical Journal. Oct 1996;**319**(2):591-599

[69] Sivamani E, DeLong RK, Qu R. Protamine-mediated DNA coating remarkably improves bombardment transformation efficiency in plant cells. Plant Cell Reports. Feb 2009; **28**(2):213-221

[70] Chu Y, Bhattacharya A, Wu C, Knoll JE, Ozias-Akins P. Improvement of peanut (*Arachis hypogaea* L.) transformation efficiency and determination of transgene copy number by relative quantitative real-time PCR. In Vitro Cellular & Developmental Biology-Plant. Jun 2013;**49**(3):266-275

[71] Lowe BA, Shiva Prakash N, Way M, Mann MT, Spencer TM, Boddupalli RS. Enhanced single copy integration events in corn via particle bombardment using low quantities of DNA. Transgenic Research. Dec 2009;**18**(6):831-840

[72] Lazzeri PA, Brettschneider R, Lührs R, Lörz H. Stable transformation of barley via PEG-induced direct DNA uptake into protoplasts. Theoretical and Applied Genetics. Apr 1991;**81**(4):437-444

[73] Caboche M. Liposome-mediated transfer of nucleic acids in plant protoplasts. Physiologia Plantarum. May 1990;**79**(1):173-176

[74] Gelvin SB. Agricultural biotechnology: Gene exchange by design. Nature. Feb 2005; **433**(7026):583-584

[75] Xiong Y, Jung J, Zeng Q, Gallo M, Altpeter F. Comparison of procedures for DNA coating of micro-carriers in the transient and stable biolistic transformation of sugarcane. Plant Cell, Tissue and Organ Culture. Jan 2013;**112**(1):95-99

[76] Tizaoui K, Kchouk ME. Genetic approaches for studying transgene inheritance and genetic recombination in three successive generations of transformed tobacco. Genetics and Molecular Biology. Jul 2012;**35**(3):640-649

[77] Yin Z, Plader W, Malepszy S. Transgene inheritance in plants. Journal of Applied Genetics. 2004;**45**(2):127-144

[78] Naylor LH. Reporter gene technology: The future looks bright. Biochemical Pharmacology. Sep 1999;**58**(5):749-757

[79] Dai S, Zheng P, Marmey P, Zhang S, Tian W, Chen S, et al. Comparative analysis of transgenic rice plants obtained by agrobacterium-mediated transformation and particle bombardment. Molecular Breeding. Jan 2001;**7**(1):25-33

[80] Gheysen G, Montagu MV, Zambryski P. Integration of *Agrobacterium tumefaciens* transfer DNA (T-DNA) involves rearrangements of target plant DNA sequences. Proceedings of the National Academy of Sciences of the United States of America. Sep 1987;**84**(17):6169-6173

[81] Bubner B, Baldwin IT. Use of real-time PCR for determining copy number and zygosity in transgenic plants. Plant Cell Reports. Nov 2004;**23**(5):263-271

[82] Park D, Park SH, Ban YW, Kim YS, Park KC, Kim NS, Kim JK, Choi IY. A bioinformatics approach for identifying transgene insertion sites using whole genome sequencing data. BMC Biotechnology. 2017;**17**(1):67

[83] Elbaidouri M, Chaparro C, Panaud O. Use of next generation sequencing (NGS) technologies for the genome-wide detection of transposition. In Plant Transposable Elements. 2013;**1057**:265-274

[84] Lambirth KC, Whaley AM, Schlueter JA, Bost KL, Piller KJ. Contrails: A tool for rapid identification of transgene integration sites in complex, repetitive genomes using low-coverage paired-end sequencing. Genomics Data. 2015;**6**:175-181

[85] Uzogara SG. The impact of genetic modification of human foods in the 21st century: A review. Biotechnology Advances. May 2000;**18**(3):179-206

[86] Rischer H, Oksman-Caldentey KM. Unintended effects in genetically modified crops: Revealed by metabolomics? Trends in Biotechnology. 2006;**24**(3):102-104

[87] Gaj T, Gersbach CA, Barbas CF. ZFN, TALEN, and CRISPR/Cas-based methods for genome engineering. Trends in Biotechnology. Jul 2013;**31**(7):397-405

Section 2

Plant Omics

Chapter 4

Plant Metabolomics: An Emerging Technology for Crop Improvement

Kapil Sharma, Supriya Sarma, Abhishek Bohra, Abhijit Mitra, Naveen K. Sharma and Anirudh Kumar

Additional information is available at the end of the chapter

http://dx.doi.org/10.5772/intechopen.76759

Abstract

The astounding ability of plants to make smart decisions in response to environment is evident. As they have evolved a long list of complex and unique processes that involve photosynthesis, totipotency, long-distance signaling, and ability to restore structural and metabolic memory, recognition, and communication via emission of the selected class of volatiles. In recent years, use of metabolite profiling techniques in detection, unambiguous identification, quantification, and rapid analysis of the minute quantity of cellular micromolecules has increased considerably. Metabolomics is key to understand the chemical footprints during different phases of growth and development of plants. To feed the ever-increasing population with limited inputs and in a rapidly changing environment is the biggest challenges that the world agriculture faces today. To achieve the project genetic gains, the breeding strategies employing marker-assisted selection for high-yielding varieties and identifying germplasm resistant to abiotic and biotic stresses are already in vogue. Henceforth, new approaches are needed to discover and deploy agronomically important gene/s that can help crops better withstand weather extremes and growing pest prevalence worldwide. In this context, metabolic engineering technology looks viable option, with immense potential to deliver the future crops.

Keywords: metabolomics, mass spectroscopy, metabolic engineering, crops, breeding

1. Introduction

Metabolomics is one of the fascinating disciplines in '– omics' field involving plants, animals, and microorganisms. Since its adoption in the mid-1990s in the field of plant biology, this approach has been successfully used in identifying important gene(s) in plants [1, 2]. The model plant *Arabidopsis thaliana* (henceforth referred to as Arabidopsis) has been extensively researched using a plethora of genomic tools and technologies, facilitating functional genomics analyses. In recent years, metabolomics approach has been extended in crop plants to ascertain gene functions [3, 4]. The ability of metabolome to serve as an ultimate phenotype of a cell renders it immensely promising for advancing crop-breeding gains [5]. For instance, delineating metabolite quantitative loci (mQTL) in crop plants offers information about the genomic target regions or genes that hold great relevance to breeding [6, 7]. Also, food and agronomical traits of crops improved through genetic modification (GM) could be better evaluated in terms of the metabolites present [8, 9].

During the last decades, techniques used to analyze metabolites have shown unprecedented refinements such as improvements in mass spectrometry (MS) and nuclear magnetic resonance spectroscopy (NMR), in conjunction with the growing ability of bioinformatics. In this chapter, we present the application of metabolomics for functional genomics in crops as well as its possible integration with crop breeding to deliver future crops.

2. Different platforms to gather metabolomic data

Let us take an example of tomato as a model system that contains different categories of chemical compounds contributing to the fruit quality. These include sugars, organic acids, amino acids, fatty acids, isoprenoids, and polyphenolic compounds. Variety of separation approaches have been used to investigate the tomato metabolome, using both targeted and nontargeted metabolomics, leading to a wide range of quality biomarkers. Targeted metabolomics is by far the most common way, as most research programs focused on understanding or improving a single target trait. A great deal of information exists that explain the phenotypic variation; however, this information may not be easily accessible.

Small molecules can have large effects. For example, the variation in the ratio between sweetness and acidity causes tomatoes to taste sharp, sweet, insipid, or lovely [10]. Accelerating improvements through breeding programs demands large-scale and low-cost assays that allow analysis of thousands of samples within a short period of time [11]. Phenotypic surveys of diverse germplasm have a very broad scope and help defining the range of acceptable phenotypic variation, albeit limited in their depth. These kinds of data on organic acid and sugar can be leveraged with gene expression analysis for discovering the genetic causes underlying fruit quality [12]. The information on carbohydrates and organic acids can also be obtained using more sophisticated tools such as nuclear magnetic resonance (NMR) spectroscopy, which detects more compounds per assay than enzymatic or colorimetric methods but at far lower throughput [13]. NMR spectroscopy is used for structural determination of a novel metabolite of particular interest. Alternatively, gas chromatography (GC) paired with mass spectrometry (MS) (GC–MS) permits broad-scope metabolomic profiling, with increased throughput compared to the NMR [14]. On the flip side, the need of GC–MS for chemical derivatization may cause exclusion of some metabolites from the analysis, and also may not produce sufficient information for the clear identification of a particular metabolite. However, combining multiple datasets emanating from complementary analytical platforms offers a powerful strategy to analyze metabolomes.

In tomato, color and aroma are other targets for improvement. A majority of pigments in tomato are isoprenoids, such as carotenoids, while others are polyphenolics (e.g., flavonoids) [15]. Traditionally, liquid chromatography (LC) with commercial standards is used for carotenoid profiling [16]. However, LC–MS is to be used for more complete estimate of metabolomes especially for isoprenoids. The MS analysis is done either inline with the LC or in an offline mode [17, 18]. Inline MS simplifies work flow, while offline MS may enhance sensitivity due to the greater reduction of sample complexity [18]. NMR spectroscopy could also be for isoprenoid profiling, which is effective in distinguishing *E* and *Z* isomers; not possible from MS analysis [19]. This is important as different carotenoid isomers may have different biological activities, hence, nutritive qualities [20]. Carotenoid composition may change during food preparation and processing, both in quality (i.e., isomerization) and identity (i.e., degradation by heat). Therefore, analysis of both raw and cooked samples is necessary for complete description of the isoprenoids [21, 22]. In addition to color, carotenoids also contribute to fruit aroma, as do fatty acid and amino acid derivatives [23]. All three represent volatile compounds, GC and GC–MS are used for their separation and identification [23, 24]. A metabolite survey of approximately 100 Dutch tomato cultivars was conducted using LC–MS and MS/MS [25].

Need for a highly curated database is one of the challenges routinely faced while analyzing MS or NMR data in order to better understand the spectra produced during an experiment. Fortunately, recent developments in tomato metabolomics have led to creation of such community-oriented resources.

In recent past, several software and analyzing tools has been developed for processing and analyze the metabolite data but till now none of the platform is self-sufficient to fulfill the user expectations. In this context, Department of Biotechnology, Government of India, has initiated a project to develop a platform (Computational Core for Plant Metabolomics, CCPM) that is a web-based collaborative platform for researchers in the field of metabolomics to store, analyze, and share their data [26].

3. Gene identification

Metabolomics study helps identifying particular mQTL which corresponds to gene(s) related to that particular trait. The method is increasingly gaining recognition because once mQTL is identified then it became easier to pin-point gene(s) responsible for that particular metabolite [27].

4. Breeding program

Researchers/breeders are interested in selecting desirable genotypes from a large plant population. Initial selection procedures relied solely on the phenotypic appearance of the plants but information on the entire breeding cycle is required (a time of nearly 10 years) to release an improved variety. To reduce this time duration, marker-based technologies such as enzyme-based markers, marker-assisted selection (MAS), and so on have been employed, that shortened the entire process up to 6 years. By using mQTL-based selection, we may further reduce time up to 4 years, given the fact that most of the metabolites are directly related to particular phenotype; and selection of mQTL remains easier and faster than that of MAS [28].

5. Metabolomic approaches to improve rice quality

Rice is an important staple crop worldwide. The crop has been benefitted considerably from the developments in the field of genomics. For example, rice genome has been sequenced and is found to encode approximately 32,000 genes [29]. However, the biological functions of more than half of these genes are yet to be determined [30]. Novel genes in rice have been identified using gain and loss-of-function approaches. Genetic linkage and association analyses with genetic core collections and segregating populations have been employed to investigate the direct relationships between metabolic composition, genotypes, and phenotypes as representatives for agronomical traits. These strategies can also be applied for other crops and vegetables (**Figure 1**). In the following section, we shall describe some of these approaches.

5.1. Approaches to collate metabolite, phenotypic, and genotypic data: some examples in rice are as follows

5.1.1. Gain-of-function approach

Construction of the rice full-length (FL) cDNA collection (*Oryza sativa* L. ssp. *japonica* "Nipponbare") was possible due to the development of the FOX hunting system (FL-cDNA

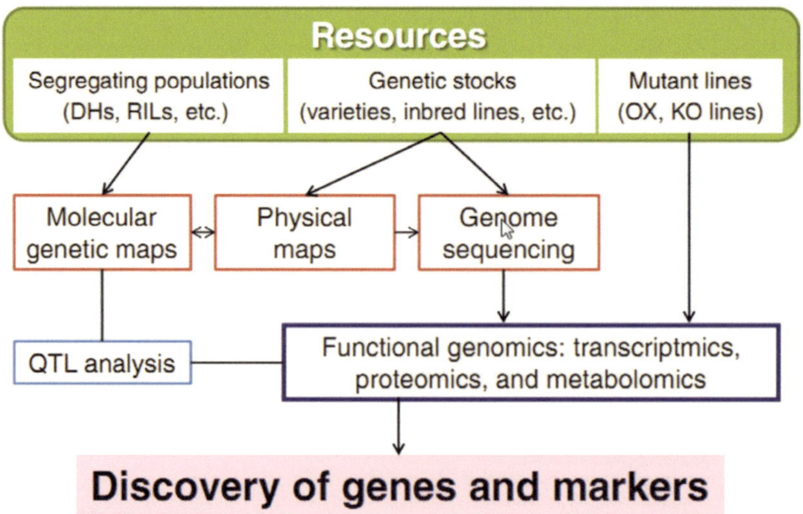

Figure 1. An overview of gene discovery and markers for crop improvement based on genetic and genomic strategies [31].

overexpressor gene hunting system) [32]. The FOX hunting system is unique, as it permits ectopic expression of any plant FL-cDNA library even in heterologous plant systems, therefore, allowing the functional analysis of genes. More than 30,000 transgenic *Arabidopsis* lines overexpressing rice FL-cDNAs, called "rice FOX *Arabidopsis* lines," have been generated [33]. Metabolic fingerprinting [34] and metabolic profiling [35] have been used with these FOX lines to identify functional genes in rice.

To screen a large number of rice FOX *Arabidopsis* lines, a nondestructive analytical method was developed using Fourier transform-near-infrared (FT-NIR) spectroscopy [34]. Unlike MS techniques, FT-NIR analysis circumvents destructive preparation, and allows data acquisition within a very short span of time (<1 min). The authors analyzed approximately 3000 FOX seeds with FT-NIR to obtain their metabolite fingerprints. Assessment of the changes in the metabolite fingerprints of the re-transformants led the discovery of seven lines with altered metabolite fingerprints in seeds. Five of these seven lines have annotations for inserted FL-cDNAs. The association of the genes with biological processes highlighted the role of complex networks underlying metabolomic responses in plants.

A detailed metabolite composition can be obtained in non-targeted manner by using metabolite profiling based on gas chromatography-time-of-flight-MS (GC-TOF-MS), particularly for primary metabolites and intermediates of secondary metabolites [36]. A set of 26 candidate lines for gene characterization were identified through surveying 350 rice FOX *Arabidopsis* lines with GC-TOF-MS. These candidate lines included a rice FOX *Arabidopsis* line that overexpressed the FL-cDNA of the rice *Lateral Organ Boundaries (LOB) Domain (LBD)/Asymmetric Leaves2-like* (*ASL)LBD37/ASL39* (Os-*LBD37/ASL39)* gene, which showed significant changes in nitrogen metabolism in the mutants [35]. The aerial parts of the rice FOX *Arabidopsis* plants exhibited hyponastic leaves and early flowering. The *Arabidopsis* At-LBD37/ASL39-overexpressor plants showed similar morphological leaf changes (i.e., hyponastic leaves), and had increased levels of amino acids and metabolites related to nitrogen metabolism. Subsequent profiling of metabolites and transcriptomes of the rice Os-*LBD37/ASL39*-overexpressing lines ascertained the same function of Os-*LBD37/ASL39* in rice and *Arabidopsis*. The analysis revealed notable features in rice overexpressor plants including early heading, metabolite alterations (related to nitrogen metabolism), and advanced leaf senescence. These findings established a close association between Os-*LBD37/ASL39* and nitrogen metabolism in rice.

Above studies suggest that the FOX hunting system can quickly and efficiently identify and characterize the genes from available cDNA libraries; the alterations that exert influence on metabolite profiles in crops and vegetables.

5.1.2. Loss-of-function approach

The *Tos17* retrotransposon- and *Ds*-transposon-inserted mutant lines have served as loss-of-function resources for characterization of the novel genes in rice [37, 38]. *Tos17*-knockout lines characterized glutamine synthetase (GS), catalyzes the key step of ammonium assimilation. Tabuchi et al. (2005) used the *Tos17*-retrotransposon inserted lines to show that the three genes (*OsGS1;1*, *OsGS1;2*, and *OsGS1;3*) encoding cytosolic GS (GS1) in rice. The *OsGS1;1* gene was critical for normal growth and grain filling [39]. They further investigated the metabolomic changes and metabolite-to-metabolite correlations of the mutants by a GC-TOF-MS-based assay [40]. In comparison to the wild-type rice, the mutants showed dramatic increase in the levels of sugars and sugar phosphates and reduced levels of amino acids and rice leaf TCA cycle intermediates. Changes in the metabolite profiles differed in root and leaf parts in the presence of ammonium. Interestingly, an overabundance was noted for nitrogen-containing secondary metabolites. The study uncovered new correlations between the over-accumulated metabolites and some primary metabolites in the mutant roots. These findings demonstrated OsGS1;1 playing crucial role in regulating the global metabolic network in rice plants grown using ammonium as the nitrogen source.

5.2. Association analysis between trait and metabolites

Modern crop-breeding practices have been highly successful in improving some important traits, for example, field performance and yield. However, genetic bottlenecks develop due to slow selection processes and narrow genetic base. Strategies to determine relationships between metabolic composition and genotypes and phenotypes in rice are discussed later.

5.2.1. Untargeted high-coverage metabolomic characterization of the rice diversity research set (RDRS)

The vast reservoir of rice seed banks provides a rich opportunity to identify genotypes possessing useful agronomical traits. However, large-scale characterization of this vast germplasm demands considerable time and resources. As a result, genetic core collections have been developed as a manageable representation of the genetic diversity. Examples include, the rice diversity research set (RDRS) comprising 67 varieties, created with the analysis of 332 varieties of *O. sativa* using restriction fragment length polymorphism (RFLP) marker [41]. To investigate the direct relationship between metabolite [5] and phenotype in RDRS, untargeted high-coverage metabolomic characterization and constructed was performed, leading to the development of predictive metabolome-trait models using multivariate regression analysis [42]. Combined datasets of rice kernels were obtained from four types of MS platforms: GC-TOF-MS for small compounds, including primary metabolites; ultra-pressure liquid chromatography-quadruple-TOF-MS (UPLC-Q-TOF-MS) for hydrophilic compounds; capillary electrophoresis-TOF-MS (CE-TOF-MS) for ionic compounds; and liquid chromatography-ion trap-TOF-MS (LC-IT-TOF-MS) for polar lipids. The study precisely defined a correlation between genetic diversity and metabolite abundance [43]. After the removal of covariance between the trait data and the population membership, a multi-block-orthogonal projection was conducted for latent structures (MB-OPLS) regression analysis. Traits such as amylose/total starch ratio and ear emergence day can be predicted from the metabolic composition by using the MB-OPLS model. The model for the amylose/total starch ratio showed a tight and negative correlation with fatty acids and lysophosphatidylcholines (**Figure 2**). Evaluation of the model using an external set of RDRS samples, other rice varieties, and the two mutants, showed high-, middle-, and low-amylose/total starch ratios, respectively. The amylose/total starch ratio was found to be associated with metabolites in rice kernels of the cultivars. However, this association was not observed in the mutants. The two loss-of-function mutants-*e1, a starch synthase IIIa* (*SSIIIa*)-deficient mutant and the *SSIIIa/starch branching enzyme* (*BE*) double-knockout mutant 4019—showed a high amylose/total starch ratio [42, 44]. Examination of starch granules with scanning electron microscopy (SEM) showed that the starch granules of the mutants were loosely packed in rice kernels [45]. Thus, fatty acids and lysophosphatidylcholines most likely play a role in packing normal starch granules into rice kernels.

5.2.2. mQTL analysis using back-cross inbred (BIL) lines

Matsuda et al. (2012) investigated 85 BILs generated by backcrossing *O. sativa* L. ssp. *japonica* "Sasanishiki" and *O. sativa* L. ssp. *indica* "Habataki" to find an association between genotype and metabolic composition [6]. The genotypic data recorded on such mapping populations are useful for QTL mapping of various agronomical traits. The genotypic data of the BIL lines cover 12 rice chromosomes, and the genotype of each BIL line was analyzed with 236 RFLPs [46]. A metabolite profiling using multi-MS-based pipelines yielded a metabolite profile dataset comprising 759 metabolite signals. Of these, 131 metabolites were identified or annotated. The lower heritability of the mQTL in yeast, mice, humans, and *Arabidopsis* than that of the expression QTL (eQTL) [47, 48] could be attributable to greater susceptibility of metabolite accumulation to environmental factors [4]. Therefore, they evaluated the effects of heritable factors on the 759 metabolic traits. Although more than half of the metabolic traits showed relatively low

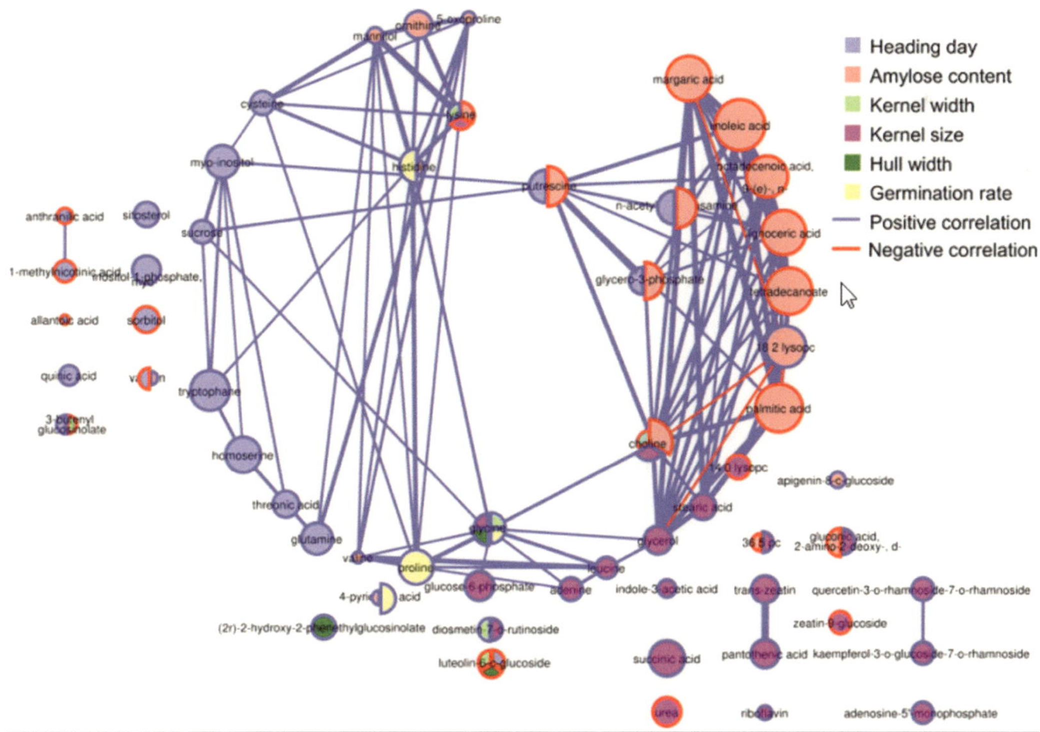

Figure 2. Correlation network of trait-associated metabolites. The node color indicates the associated trait. Red lines (edges) represent positive correlations, while purple edges show negative correlations. The thickness of the edges indicates the strength of the correlation [31].

broad-sense heritability (H^2), high H^2 values were observed for some of the secondary metabolites, such as lysophosphatidylcholines, oryzanols, and flavone glycosides. Notably, heritability profiles obtained in rice were not similar to those of tomato fruits and *Arabidopsis* leaves [49, 50]. The QTL mapping results identified 802 mQTL from 759 metabolic traits and suggested for a coordinated control of some metabolites, such as amino acids and triacylglycerols, through a mQTL hotspot on chromosome three. The extent of genetic control was determined for the annotated flavone glycoside level. The authors determined the structure of the flavone glycoside by using multi-step chromatography, MS, and NMR. The mQTL analysis provides faster and efficient breeding technique to dissect useful metabolic traits of both primary and secondary metabolites in rice.

6. Metabolomic approach to improve legume crops

Forage and grain legumes contribute 27% of the world gross primary crop. The grain legumes alone cater 33% of required human dietary protein, thus contributing to the global food security and environmental sustainability [51, 52]. Barring a few extensively investigated model legumes, metabolomics studies in other legumes remain limited. The studies in model legumes demonstrate a decrease in oxylipins as effect of rhizobial node factor (Nod) in *Medicago* [53] and metabolic adjustments of shoot constituent in salt tolerant *Lotus* species for its survival [54].

Stress conditions such as salinity and anoxia cause an accumulation of alanine, and its biosynthesis co-substrates such as glutamate and GABA, and succinate in soybean [55]. Differential expression was also obtained for genes involved in nitrogen fixation and fermentation in root. Interestingly, a negative correlation was observed for amino acid derived from glycolysis and the TCA cycle during water logging; several TCA cycle enzymes were induced upon exposure to water logging [56]. Likewise, a study on metabolic changes associated with flooding stress in soybean revealed a set of 81 mitochondria-associated metabolites, suggesting a boost in concentrations of metabolites involved in respiration and glycolysis such as, amino acids, NAD, and NADH coupled with the depletion of free adenosine triphosphate (ATP) [57]. Under drought and salinity conditions, metabolite phenotyping of four different Mediterranean accessions of lentil suggested a decrease in intermediates of the TCA cycle and glycolytic pathway [58]. Importantly, the study yielded metabolite markers for specific stress; such as threonate, asparagine/ornithine, and alanine/homoserine for NaCl, drought, and salinity, respectively. Another study aimed to assess the impact of water deficiency on *Lupinus albus* demonstrated that the plant stem served as a storage organ for sugars and amino acids [59]. Importantly, tolerant plant accumulated high level of metabolites such as asparagine, proline, sucrose, and glucose in the stem stelar region [59]. This suggests for reorganization of nitrogen and carbon metabolism pathways in plants in order to tolerate salinity stress. In soybean, consistent increase in pinitol (sugar alcohol, osmoprotectant) was reported in the tolerant plant at both normal and drought-stressed conditions [60]. Similarly, accumulation of sucrose, free amino acids, and soluble proteins was observed in tolerant soybean in response to water stress [61].

7. Metabolomic approaches to evaluate GM crops

GM crops are now widely used worldwide [62]. The International Service for the Acquisition of Agri-Biotech Applications (ISAAA) reported that in 2011, 160 million hectares of arable land was used to grow biotech crops, including GM crops (http://www.isaaa.org/).

Metabolism refers to the processes involved in maintaining life, such as the synthesis and breakdown of proteins, nucleic acids, and carbohydrates. Metabolomics offers a snapshot of the current biochemical status, including important nutritional and toxicological characteristics. Furthermore, the metabolite composition is reported to have close association with the organism's phenotype. Hence, metabolomics is a useful tool for investigating the metabolic composition of GM crops. The application of metabolomic technology could generate a database of metabolites in both GM crops and traditional varieties. For instance, metabolomics approach was employed to assess the chemical composition of GM tomatoes in order to compare the modified crops with the traditional varieties [63]. The authors used GM tomatoes overexpressing a foreign gene encoding miraculin, a glycoprotein found in tropical plants but normally absent in tomatoes [64]. The MS-based multiple platforms detected 86% of the total chemical diversity in the tomato cultivars used in the study. Subsequently, statistical approach for "proof-of-safety" rather than "proof-of hazard" approach was used to evaluate "similarities" and "differences" between GM tomatoes and six traditional cultivars, including the control line Moneymaker. Results suggested that the GM tomatoes had a reproducible metabolic signature; moreover, more than 92% of the compounds showed an acceptable variation in both green and red stages of the tomato, highlighting striking similarity of the GM tomatoes with that of the control line Moneymaker in terms of their metabolite profiles.

Furthermore, a comparison was drawn for the metabolite profiles obtained from two independent experiments. The study determined the levels of the most commonly altered metabolites in the GM tomatoes, such as proline, 4-hydroxy-proline, spermidine, asparagine, arginine, serine, and inositol-1-phosphate, across all growth conditions. The expression of these metabolites was unaltered by genetic modification, not associated with the expression of foreign genes. This approach could be useful for evaluating GM crops for assessing their metabolomic equivalence with traditional crops.

8. Conclusions and future perspective

The growing attention that metabolomics is receiving in the field of plant research could be ascribed to plant's ability to produce a vast array of metabolites, far greater than that produced by animals and microorganisms. Achieving a comprehensive coverage of metabolome analysis calls for multiparallel complementary technologies instead of relying on a single analytical technology. Increasing the annotation rate of unknown signals still poses a big challenge. The cooccurrence principle of transcripts and metabolites, particularly transcriptome co-expression network analysis, is powerful for decoding functions of genes not only in a model plants but also in crops and medicinal plants. The mQTL analysis along with scoring of gene expression and agronomical traits emerges as a promising technique to support crop breeding [65]. In addition to expedite the development of improved cultivars, metabolomics plays a key role in the evaluation of GM crops.

Combining de novo transcriptome assembly [66] and metabolomic techniques enables us to adopt a systems biology approach to investigate genetic populations as both techniques do not require a reference genome sequence. These post-genomics tools and techniques can considerably shorten the time required for selection in plant breeding and accelerate the discovery of novel genes in crops, vegetables, and medicinal plants [67, 68]. In summary, systems biology, metabolomics, and other omics will play a key role in understanding plant systems and developing novel biotechnology applications for crop improvement.

Acknowledgements

The authors are grateful to DBT, India for funding to Computational Core for Plant Metabolomics (CCPM) project jointly at IIIT-Hyderabad and JNU-Delhi, (No. BT/PR14715/PBD/16/903/2010). Authors are grateful to thank Prof. Indira Ghosh for her consistent guidance and support. K.S and S.S are grateful to DBT for the fellowship support.

Author details

Kapil Sharma[1], Supriya Sarma[2], Abhishek Bohra[3], Abhijit Mitra[1], Naveen K. Sharma[4] and Anirudh Kumar[4]*

*Address all correspondence to: anirudh.kumar@igntu.ac.in

1 Center for Computational Natural Sciences and Bioinformatics (CCNSB), IIIT-Hyderabad, India

2 CSIR-Centre for Cellular & Molecular Biology (CCMB), Hyderabad, India

3 Crop Improvement Division, Indian Institute of Pulses Research (IIPR), Kanpur, India

4 Department of Botany, Indira Gandhi National Tribal University (IGNTU), Amarkantak, India

References

[1] Bino RJ, Hall RD, Fiehn O, Kopka J, Saito K, Draper J, Nikolau BJ, et al. Potential of metabolomics as a functional genomics tool. Trends in Plant Science. 2004;**9**:418-425

[2] Saito K, Matsuda F. Metabolomics for functional genomics, systems biology, and biotechnology. Annual Review of Plant Biology. 2010;**61**:463-489

[3] Oikawa A, Matsuda F, Kusano M, Okazaki Y, Saito K. Rice. Rice metabolomics. 2008;**1**:63-71

[4] Fernie AR, Schauer N. Metabolomics-assisted breeding: A viable option for crop improvement? Trends in Genetics. 2009;**25**:39-48

[5] Hall RD. Plant metabolomics: From holistic hope, to hype, to hot topic. The New Phytologist. 2006;**169**:453-468

[6] Matsuda F, Okazaki Y, Oikawa A, Kusano M, Nakabayashi R, et al. Dissection of genotype-phenotype associations in rice grains using metabolome quantitative trait loci analysis. The Plant Journal. 2012;**70**:624-636

[7] Okazaki Y, Saito K. Recent advances of metabolomics in plant biotechnology. Plant Biotechnology Reports. 2012;**6**:1-15

[8] Catchpole GS, Beckmann M, Enot DP, Mondhe M, Zywicki B, Taylor J, et al. Hierarchical metabolomics demonstrates substantial com-positional similarity between genetically modified and conventional potato crops. Proceedings of the National Academy of Sciences of the United States of America. 2005;**102**:14458-14462

[9] Baker JM, Hawkins ND, Ward JL, Lovegrove A, Napier JA, Shewry PR, Beale MH. A metabolomic study of substantial equivalence of field grown genetically modified wheat. Plant Biotechnology Journal. 2006;**4**:381-392

[10] Baldwin EA, Scott JW, Shewmaker CK, Schuch W. Flavor trivia and tomato aroma: Biochemistry and possible mechanisms for control of important aroma components. Hort Science. 2000;**35**:1013-1022

[11] Velterop JS, Vos F. A rapid and inexpensive microplate assay for the enzymatic determination of glucose, fructose, sucrose, L-malate and citrate in tomato (*Lycopersicon esculentum*) extracts and in orange juice. Phytochemical Analysis. 2001;**12**(5):299-304

[12] Baxter CJ, Carrari F, Bauke A, et al. Fruit carbohydrate metabolism in an introgression line of tomato with increased fruit soluble solids. Plant & Cell Physiology. 2005;**46**(3):425-437

[13] Sobolev AP, Segre A, Lamanna R. Proton high-field NMR study of tomato juice. Magnetic Resonance in Chemistry. 2003;**41**:237-245

[14] Fraser PD, Enfissi EM, Halket JM, et al. Manipulation of phytoene levels in tomato fruit: Effects on isoprenoids, plastids, and intermediary metabolism. The Plant Cell. 2007;**19**(10):3194-3211

[15] Grotewold E. The genetics and biochemistry of floral pigments. Annual Review of Plant Biology. 2006;**57**:761-780

[16] Enfissi EM, Fraser PD, Lois LM, Boronat A, Schuch W, Bram-ley PM. Metabolic engineering of the mevalonate and non-mevalonate isopentenyl diphosphate-forming pathways for the production of health-promoting isoprenoids in tomato. Plant Biotechnology Journal. 2005;**3**(1):17-27

[17] Capanoglu E, Beekwilder J, Boyacioglu D, Hall R, de Vos CH. Changes in antioxidant and metabolite profiles during production of tomato paste. Journal of Agricultural and Food Chemistry. 2008;**56**(3):964-973

[18] Fraser PD, Enfissi EM, Goodfellow M, Eguchi T, Bramley PM. Metabolite profiling of plant carotenoids using the matrix-assisted laser desorption ionization time-of-flight mass spectrometry. The Plant Journal. 2007;**49**(3):552-564

[19] Tiziani S, Schwartz SJ, Vodovotz Y. Profiling of carotenoids in tomato juice by one- and two-dimensional NMR. Journal of Agricultural and Food Chemistry. 2006;**54**(16):6094-6100

[20] Yeum KJ, Russell RM. Carotenoid bioavailability and bioconversion. Annual Review of Nutrition. 2002;**22**:483-504

[21] Dewanto V, Wu X, Adom KK, Liu RH. Thermal processing enhances the nutritional value of tomatoes by increasing total antioxidant activity. Journal of Agricultural and Food Chemistry. 2002;**50**(10):3010-3014

[22] Re R, Bramley PM, Rice-Evans C. Effects of food processing on flavonoids and lycopene status in a Mediterranean tomato variety. Free Radical Research. 2002;**36**(7):803-810

[23] Tieman DM, Zeigler M, Schmelz EA, et al. Identification of loci affecting flavour volatile emissions in tomato fruits. Journal of Experimental Botany. 2006;**57**(4):887-896

[24] Tikunov Y, Lommen A, de Vos CH, et al. A novel approach for nontargeted data analysis for metabolomics. Large-scale profiling of tomato fruit volatiles. Plant Physiology. 2005;**139**(3):1125-1137

[25] Moco S, Bino RJ, Vorst O, et al. A liquid chromatography-mass spectrometry-based metabolome database for tomato. Plant Physiology. 2006;**141**(4):1205-1218

[26] Pudi V, Rani P, Mitra A, Ghosh I. Computational core for plant metabolomics: A case for interdisciplinary research. In: Reddy P, Sureka A, Chakravarthy S, Bhalla S, editors. Big Data Analytics. BDA 2017. Lecture Notes in Computer Science, Vol. 10721. Cham: Springer

[27] Lisec J, Meyer RC, Steinfath M, Redestig H, Becher M, et al. Identification of metabolic and biomass QTL in *Arabidopsis thalianaina* parallel analysis of RIL and IL populations. The Plant Journal. 2008;**53**:960-967

[28] Alisdair R. Fernie and Nicolas Schauer. Metabolomics-assisted breeding: A viable option for crop improvement? Trends in Genetics. 2008;**25**:39-48

[29] Sato Y, Antonio BA, Namiki N, Takehisa H, Minami H, Kamatsuki K, et al. RiceXPro: A platform for monitoring gene expression in japonica rice grown under natural field conditions. Nucleic Acids Research. 2011;**39**:D1141-D1148

[30] Itoh T, Tanaka T, Barrero RA, Yamasaki C, Fujii Y, Hilton PB, et al. Curated genome annotation of Oryza sativa ssp. Japonica and comparative genome analysis with Arabidopsis thaliana. Genome Research. 2007;**17**:175-183

[31] Kusano M, Saito KJ. Role of metabolomics in crop improvement. Plant Biochemistry and Biotechnology. 2012;**21**:S24–S31

[32] Ichikawa T, Nakazawa M, Kawashima M, Iizumi H, Kuroda H, et al. The FOX hunting system: An alternative gain-of-function gene hunting technique. The Plant Journal. 2006;**48**:974-985

[33] Kondou Y, Higuchi M, Takahashi S, Sakurai T, Ichikawa T, et al. Systematic approaches to using the FOX hunting system to identify useful rice genes. The Plant Journal. 2009; **57**:883-894

[34] Suzuki M, Kusano M, Takahashi H, Nakamura Y, Hayashi N, et al. Rice-*Arabidopsis* FOX line screening with FT-NIR-based fingerprinting for GC-TOF/MS-based metabolite profiling. Metabolomics. 2010;**6**:137-145

[35] Albinsky D, Kusano M, Higuchi M, Hayashi N, Kobayashi M, et al. Metabolomic screening applied to rice FOX *Arabidopsis* lines leads to the identification of a gene-changing nitrogen metabolism. Molecular Plant. 2010a;**3**:125-142

[36] Kusano M, Fukushima A, Redestig H, Saito K. Metabolomic approaches toward understanding nitrogen metabolism in plants. Journal of Experimental Botany. 2011a;**62**: 1439-1453

[37] Hirochika H, Guiderdoni E, An G, Hsing YI, Eun MY, Han CD, et al. Rice mutant resources for gene discovery. Plant Molecular Biology. 2004;**54**:325-334

[38] Kolesnik T, Szeverenyi I, Bachmann D, Kumar CS, Jiang S, Ramamoorthy R, et al. Establishing an efficient Ac/Ds tagging system in rice: Large-scale analysis of Ds flanking sequences. The Plant Journal. 2004;**37**:301-314

[39] Tabuchi M, Sugiyama K, Ishiyama K, Inoue E, Sato T, Takahashi H, Yamaya T. Severe reduction in growth rate and grain filling of rice mutants lacking OsGS1;1, a cytosolic glutamine synthetase 1;1. The Plant Journal. 2005;**42**:641-651

[40] Kusano M, Tabuchi M, Fukushima A, Funayama K, Diaz C, et al. Metabolomics data reveal a crucial role of cytosolic glutamine synthetase 1;1 in coordinating metabolic balance in rice. The Plant Journal. 2011c;**66**:456-466

[41] Kojima Y, Ebana K, Fukuoka S, Nagamine T, Kawase M. Development of an RFLP-based rice diversity research set of germplasm. Breeding Science. 2005;**55**:431-440

[42] Redestig H, Kusano M, Ebana K, Kobayashi M, Oikawa A, et al. Exploring molecular backgrounds of quality traits in rice by predictive models based on high-coverage metabolomics. BMC Systems Biology. 2011;**5**:176

[43] Mantel N. The detection of disease clustering and a generalized regression approach. Cancer Research. 1967;**27**:209-220

[44] Fujita N, Yoshida M, Kondo T, Saito K, Utsumi Y, Tokunaga T, et al. Characterization of SSIIIa-deficient mutants of rice: The function of SSIIIa and pleiotropic effects by SSIIIa deficiency in the rice endosperm. Plant Physiology. 2007;**144**:2009-2023

[45] Kusano M, Fukushima A, Fujita N, Okazaki Y, Kobayashi M, Oitome NF, Ebana K, Saito K. Deciphering starch quality of rice kernels using metabolite profiling and pedigree network analysis. Molecular Plant. 2012;**5**:442-451

[46] Nagata K, Fukuta Y, Shimizu H, Yagi T, Terao T. Quantitative trait loci for sink size and ripening traits in rice (Oryza sativa L.). Breeding Science. 2002;**52**:259-273

[47] Schadt EE, Monks SA, Drake TA, Lusis AJ, Che N, et al. Genetics of gene expression surveyed in maize, mouse and man. Nature. 2003;**422**:297-302

[48] Meyer RC, Steinfath M, Lisec J, Becher M, Witucka-Wall H, et al. The metabolic signature related to high plant growth rate in Arabidopsis thaliana. Proceedings of the National Academy of Sciences of the United States of America. 2007;**104**:4759-4764

[49] Lisec J, Meyer RC, Steinfath M, Redestig H, Becher M, Witucka-Wall H, et al. Identification of metabolic and biomass QTL in *Arabidopsis thaliana* in a parallel analysis of RIL and IL populations. The Plant Journal. 2008;**53**:960-972

[50] Schauer N, Semel Y, Balbo I, Steinfath M, Repsilber D, Selbig J, Pleban T, Zamir D, Fernie AR. Mode of inheritance of primary metabolic traits in tomato. The Plant Cell. 2008;**20**:509-523

[51] Graham PH, Vance CP. Legumes: Importance and constraints to greater use. Plant Physiology. 2003;**131**:872-877

[52] Ramalingam A, Kudapa H, Pazhamala LT, Weckwerth W, Varshney RK. Proteomics and metabolomics: Two emerging areas for legume improvement. Frontiers in Plant Science. 2015;**6**:1116

[53] Zhang N, Venkateshwaran M, Boersma M, Harms A, Howes-Podoll M, Den Os D, et al. Metabolomic profiling reveals suppression of oxylipin biosynthesis during the early stages of legume-rhizobia symbiosis. FEBS Letters. 2012;**586**:3150-3158

[54] Sanchez DH, Pieckenstain FL, Escaray F, Erban A, Kraemer U, Udvardi MK, et al. Comparative ionomics and metabolomics in extremophile and glycophytic Lotus species under salt stress challenge the metabolic pre-adaptation hypothesis. Plant, Cell & Environment. 2011;**34**:605-617

[55] Rocha M, Sodek L, Licausi F, Hameed MW, Dornelas MC, van Dongen JT. Analysis of alanine aminotransferase in various organs of soybean (*Glycine max*) and in dependence of different nitrogen fertilisers during hypoxic stress. Amino Acids. 2010b;**39**:1043-1053

[56] Rocha M, Licausi F, Araujo WL, Nunes-Nesi A, Sodek L, Fernie AR, et al. Glycolysis and the tricarboxylic acid cycle are linked by alanine aminotransferase during hypoxia induced by waterlogging of Lotus japonicus. Plant Physiology. 2010a;**152**:1501-1513

[57] Komatsu S, Yamamoto A, Nakamura T, Nouri MZ, Nanjo Y, Nishizawa K, et al. Comprehensive analysis of mitochondria in roots and hypocotyls of soybean under flooding stress using proteomics and metabolomics techniques. Journal of Proteome Research. 2011;**10**:3993-4004

[58] Muscolo A, Junker A, Klukas C, Weigelt-Fischer K, Riewe D, Altmann T. Phenotypic and metabolic responses to drought and salinity of four contrasting lentil accessions. Journal of Experimental Botany. 2015;**66**:5467-5480

[59] Pinheiro C, Passarinho JA, Ricardo CP. Effect of drought and rewatering on the metabolism of Lupinus albus organs. Journal of Plant Physiology. 2004;**161**:1203-1210

[60] Silvente S, Sobolev AP, Lara M. Metabolite adjustments in drought tolerant and sensitive soybean genotypes in response to water stress. PLoS One. 2012;**7**:e38554

[61] Tripathi P, Rabara RC, Shulaev V, Shen QJ, Rushton PJ. Understanding water-stress responses in soybean using hydroponics system-a systems biology perspective. Frontiers in Plant Science. 2015;**6**:1145

[62] Yonekura-Sakakibara K, Saito K. Review: Genetically modified plants for the promotion of human health. Biotechnology Letters. 2006;**28**:1983-1991

[63] Kusano M, Redestig H, Hirai T, Oikawa A, Matsuda F, Fukushima A, et al. Covering chemical diversity of genetically-modified tomatoes using metabolomics for objective substantial equivalence assessment. PLoS One. 2011b;**6**:e16989

[64] Sun HJ, Kataoka H, Yano M, Ezura H. Genetically stable expression of functional miraculin, a new type of alternative sweetener, in transgenic tomato plants. Plant Biotechnology Journal. 2007;**5**:768-777

[65] Wentzell AM, Rowe HC, Hansen BG, Ticconi C, Halkier BA, Kliebenstein DJ. Linking metabolic QTLs with network and cis-eQTLs controlling biosynthetic pathways. PLoS Genetics. 2007;**3**:1687-1701

[66] Martin JA, Wang Z. Next-generation transcriptome assembly. Nature Reviews. Genetics. 2011;**12**:671-682

[67] Yonekura-Sakakibara K, Saito K. Functional genomics for plant natural product biosynthesis. Natural Product Reports. 2009;**26**:1466-1487

[68] Bunsupa S, Katayama K, Ikeura E, Oikawa A, Toyooka K, Saito K, Yamazaki M. Lysine decarboxylase catalyzes the first step of quinolizidine alkaloid biosynthesis and coevolved with alkaloid production in leguminosae. The Plant Cell. 2012;**24**:1202-1216

Chapter 5

Securing Diversity for Food Security: The Case of Conservation and Use of Rice Genetic Resources

Roel C. Rabara, Marilyn C. Ferrer,
Jennifer Jara-Rabara and Rachel C. Sotto

Additional information is available at the end of the chapter

http://dx.doi.org/10.5772/intechopen.77216

Abstract

Producing enough food, fiber, and fuel, in this case, the second most important global crop called rice, remains a continuing challenge as global population increases and various production constraints ensue. Plant breeding scientists prefer using elite rice lines but also infuse new genetic resources into the parental genepool for desirable traits, such as resistance to pests and diseases, good flavor, and high nutritional quality on top of high-yielding potential. Prior research studies reveal the importance of germplasm resources including wild rice relatives as excellent sources of desirable traits in new crop breeds. Advances in molecular and genomics approaches (QTLs, GWAS, OMICs technologies) have identified and transferred genes, genomes, loci among other important genetic materials that are sought for. As knowledge builds up with these biotechniques, more rice genetic resources can be characterized at the molecular and systems levels for further utility in breeding better cultivars. Information generated from innovative approaches must be documented and processed as germplasm characterization data and must remain accessible at genebanks that exist centrally to conserve biodiversity. Development of germplasm information should be a collaborative effort of scientists who share similar interests in exploiting the valuable and novel genes within germplasm resources that are essential for crop improvement.

Keywords: rice genetic resources, diversity, molecular approaches, OMICS technologies, genomic-assisted breeding

1. Introduction

Food-production systems are greatly affected by increasing demand of a burgeoning world population on top of production constraints as such water availability, soil degradation, dwindling arable lands, and abrupt changes in climate. Commitments to uphold the fundamental right of people to have adequate food and to be free from hunger were made during the Rome Declaration on World Food Security in 1996 and the United Nations Millennium Summit in 2000. These included commitments of participating countries to implement policies on eradication of inequality and poverty through improved economic and physical access to sufficient, safe, and nutritious food supply. Despite these efforts, almost 1000 million people experience malnutrition and hunger, of which 15 million who are predominantly children die every year [1]. Producing enough food becomes more crucial considering that population is expected to grow to about 8.3 billion by 2030 where about 90% of the increase emanates from developing countries [2]. Considering such pressures, one must remember that food security is interlinked with wise use of agricultural biodiversity and genetic resources including their conservation and exchange [1].

Plant genetic resources for food and agriculture (PGRFA) are crucial for food production and sustainable agriculture as they become foundations for breeders, biotechnologists and farmers engaged in developing new plant varieties that can address growing food demands, unpredictable human needs, and changing environmental conditions [1]. Although significant to human survival, PGRFA is dwindling at an alarming rate. For example, hundreds of thousands of heterogeneous plant varieties and landraces have been developed for generations in farmers' fields until the beginning of the 20th century [1]. It is also important to note that socioeconomic perspective in agriculture varies by region. About 50% of the populations in Africa and Asia depend on agriculture, while only a meager 1.9% does in North America, and that half of the worlds' population chiefly draws income from agricultural production [2].

2. Germplasm banks and functions

Worldwide, central to conserving agricultural diversity are about 1700 genebanks including 11 international genebanks that spend about US$18 million a year [3] to maintain, manage and share germplasm collections. These 11 international genebanks are managed by the consortium of international agricultural research centers (CGIAR, formerly the Consultative Group for International Agricultural Research). Collectively, CGIAR genebanks hold 730,000 accessions in 35 collections, available as seeds, plants maintained in screenhouses or fields, in cryopreservation, in tissue culture, and as DNA samples [4]. In recent years, germplasm distribution reported an overall increase estimated at almost 40,000 samples yearly between 1985 and 2009 [5] and almost 92,000 samples yearly between 2012 and 2014 [4]. The considerable upsurge in value and demand for diversity will ensue owing to present-day advances in high-throughput sequencing and phenotyping, disease indexing, and screening data [6].

Although genebanks exist as germplasm repositories, farmers and farmers' groups, or organizations also exert efforts to save seeds of crop species and have been custodians of landraces and traditional varieties as well [7, 8]. Additionally, genebanks exist to conserve the genetic diversity of wild and cultivated plants that people rely on as a source of food, fuel, and fiber [3]. Maintaining biodiversity in both plants and animals has become a central principle in formulating strategies for sustainable agriculture advancement [7]. Diversity is important in progressing nutritional quality, productivity, and sustainability [3] of plants and animals. Diversity shown in crop wild relatives (CWRs) means beneficial traits such as resistance to biotic and abiotic stresses and adaptation to a wide range of habitats or environments [9–11], which are important attributes to curb the effects of climate change-induced variations [12] affecting crop growth and development parameters.

Throughout time, scientists and researchers including the public have witnessed and realized the long-term benefits derived from conserving and securing biological diversity [7]. The biodiversity conserved in genebanks helps advance global plant breeding programs [3]. In rice, for example, a study found out that *Oryza nivara*, a wild rice species conferred protection against grassy stunt virus to almost all tropical rice varieties in Asia [7]. Additionally, a genomics study [6] on more than 4300 rice varieties worldwide reported that 100% of the varieties from the international rice research institute (IRRI) and 90% of non-IRRI varieties have in their pedigrees at least one accession from the international rice genebank (IRG).

As of January 2017, IRRI reports that IRG maintains more than 127,916 rice accessions and 4647 wild relatives including 44 wild *Oryza* species and nine species from seven related genera [13]. This makes the Philippines as the country repository having the largest in number and most diverse rice genetic resources (**Figure 1**) as shown in Genesys, a global portal to information about PGRFA. IRG's holdings come from various rice-growing countries that transmit to IRG the seeds of rice cultivars for safeguarding and sharing as public goods [13].

Considering agricultural diversity as a public good strengthens global interdependence and proactive upkeep of plant genetic resources [14]. One of the deep-rooted forms of interdependence is the reliance of foreign genetic resources as exhibited by the spread of common

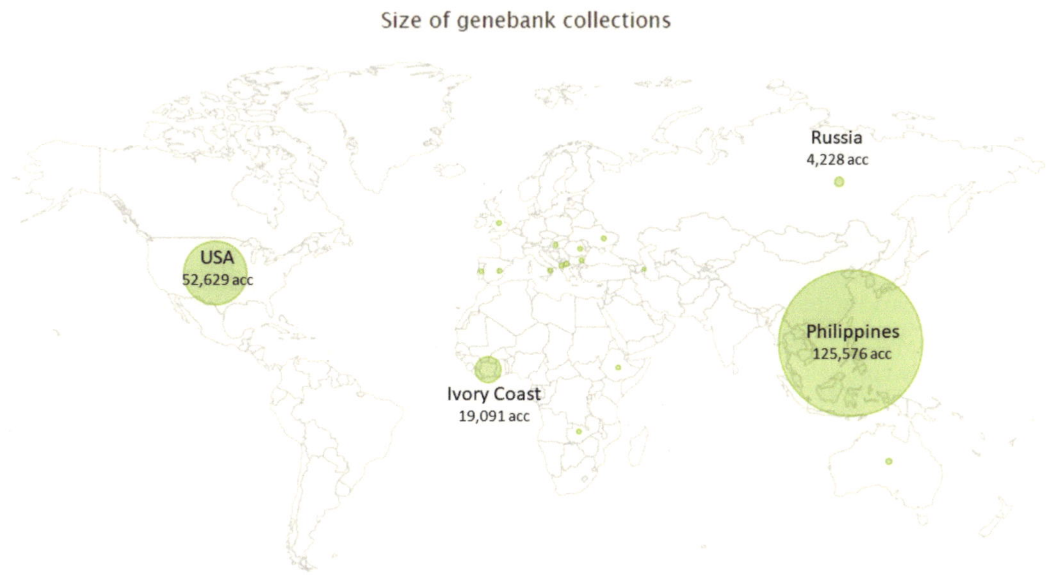

Figure 1. Genebank holdings of rice germplasm around the world (data from Genesys database https://goo.gl/xPFzsE).

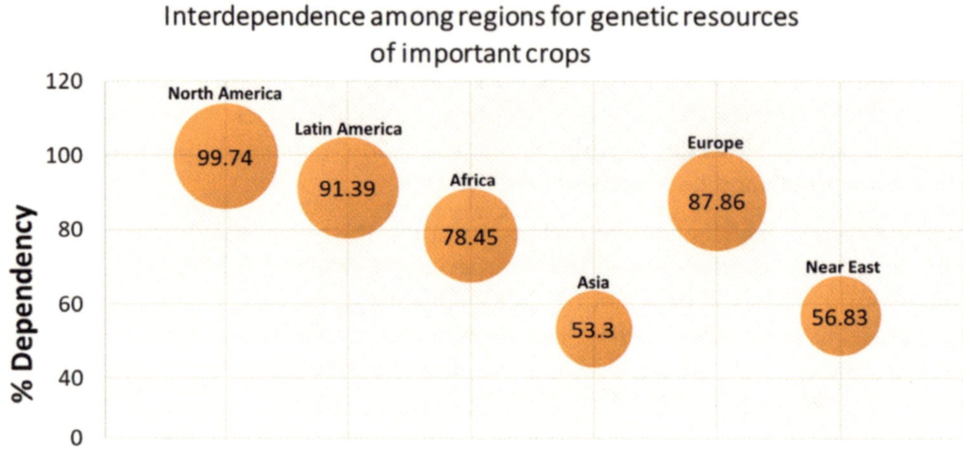

Figure 2. Estimate of the degree of interdependence of major regions on genetic resources of important crops.

crops from centers of origin to the rest of the world [1]. About 70% is the estimated average of interdependence degree of countries (**Figure 2**) on the most important crops worldwide [1]. These relate to the function of genebanks as worldwide distributors of crop germplasm to help sustain the availability of food, fiber, and fuel, strengthen on-farm crop diversity and crop productivity. The dependency of most countries and their farmers on modern, improved varieties of rice, corn, wheat, and other crops [15] existed then and likely on the rise owing to productivity concerns brought by global climate change. For example, at the top of its popularity, IR36 rice variety was planted to 10% of the world's rice area which is about 11 million hectares [16]. Moreover, modern varieties are usually created using genetic resources sourced from various countries that demonstrate the interdependence on the availability and accessibility of plant genetic resources [16] for sustainable agriculture and food security.

3. Rice production systems and constraints

Rice ranks second to maize among the most important cereals globally produced in the world [17]. Global rice production was forecasted at a record high of 481.3 million tons in 2017 with the bulk of production coming from Asia (**Figure 3A**) [18]. Rice is chiefly consumed in the developing countries (**Figure 3B**) with about 340 metric tons (Mt) and 392 Mt. (15% increase) consumed in 2002–2004 and in 2012–2014, respectively, [19] and projected to reach 449 Mt. in 2024, a 32% increase from 2004 consumption level [19]. In contrast, developed countries consumed 17 Mt. of rice in 2002–2004 and about 3% more (17.5 Mt) in 2012–2014 and the consumption is expected to reach 18.2 Mt. [19].

The South Asia region is a global major rice producer yet 75% of rice growers from this region are smallholders [20]. Smallholding farmers are key food producers in developing countries contributing about 70–80% of the food produce in Africa and in sub-Sahara and Asia, respectively [20]. Due to their major role in food production in developing countries, smallholder farmers

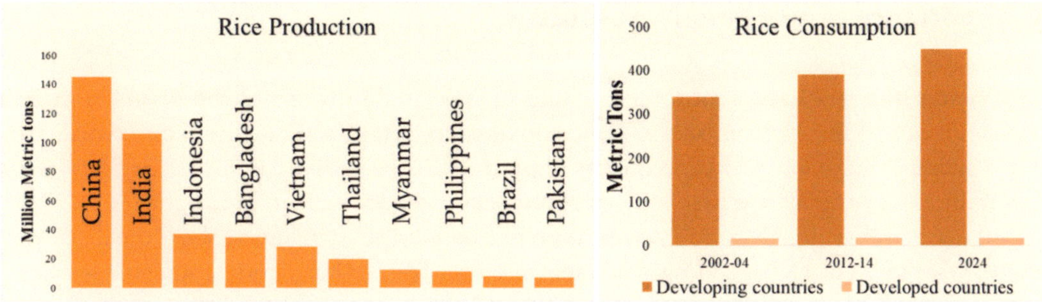

Figure 3. Rice production and consumption in the world. (A). Top 10 rice producing countries in the world. (B). Rice consumption in developing and developed countries (data from (Childs, 2017) and OECD/FAO, 2015).

are crucial contributors to food security and produce more than 50% of the food requirement to feed 9 billion people in 2050 [21]. However, these groups of farmers are highly vulnerable to production constraints, which include socio-economic, biotic, abiotic, and management-related factors (**Figure 4**) [20]. Studies have shown that 18–21% and 22–23% of yield losses are attributed to biotic and abiotic factors [20, 22]. Another major biotic factor to yield loss is weed competition contributing up to 7% yield loss in rice production of small-scale farmers [20]. Studies on no weed control in rice production incurred as high as 94–96% yield loss during the wet and dry season of 2009, respectively [23]. Insect pest that accounts 5% yield loss among smallholder farmers [20] have caused 20% loss in sub-Sahara in 2008 [24]. Socio-economic factors contributed 22% to yield loss from which 4.5% is due to difficulty in access to sufficient irrigation water [20].

Shifts in global climatic conditions also pose threats of limiting crop yields and water-supply availability among other factors that threaten the bumper food supply. In the United States, USDA-ERS reports that climate change can limit average yields in various crops including rice in both dryland and irrigated production by 2020 [25]. The projected decline in US rice production is 2.2% by 2020 to as much as 6.1% by 2080 [25].

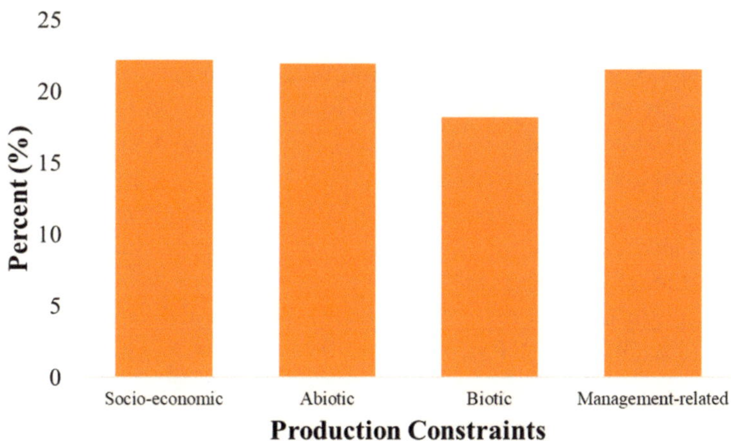

Figure 4. Major factors affecting rice production in south Asian farmers.

4. Importance of rice genetic resources

Smallholder farmers have contributed to and unknowingly preserved the diversity of rice genetic resources throughout time as they cultivated, selected, and nurture their favorite cultivars owing to favorable yield and quality of product. These cultivars comprised of landraces and traditional varieties are vast sources of important genetic traits that can further improve future rice varieties [8]. Historical accounts on rice breeding, for instance, show the discovery and development of high-yielding cultivars that can withstand environmental stresses, such as varieties having excellent genes for submergence tolerance and having significant yield increases owing to NAL1 allele [8]. One variety dubbed miracle rice, IR8 released in 1966 served as a source of desirable traits and reportedly is associated to 92% of the 67 rice varieties released in the Philippines between 1960 and 1994 and that the varieties have 57 more parental lines in commonality [26]. Moreover, 19 ancestral parents comprise the genetic core of all Philippine rice cultivars developed from 1960 to 1994 [26, 27].

Genetic diversity of the 193 rice cultivars collected from 19 countries was studied toward developing an international rice molecular program [27]. A cluster analysis using SSR data of the 193 accessions showed four prominent groups: classical *indica* (Group I), classical japonica (Group IV), and a mix of modern cultivars, traditional *japonica,* and diverse landraces (Groups II and III) of which Group I showed the highest level of diversity [27]. Among 632 alleles shown in the study, only 5% was prevalent alleles, which were consistent with the known knowledge that 95% of the rice genepool was seldom used in rice breeding programs [27].

Other notable results of the study include the possible existence of differentiation between the temperate and tropical indica that may not be exhaustive as those in japonica cultivars in same regions [27]. Additionally, the study revealed that high genetic diversity between Chinese *indica* maintainer lines and the tropical restorer line developed at the Philippine-based IRRI highly influenced the hybrid vigor of Chinese hybrids [27]. The study further suggested that molecular breeding of *japonica-indica* variation be explored to further increase the genetic diversity of rice cultivars that may improve production sustainability and help break the yield ceiling [27].

A recent genomics research study underscores the importance of genetic conservation in rice. Using 13 reference genomes of modern and related-rice species, the study found rapid species diversification reflected by lineage-specific emergence and turnover of various novel elements, consisting of potential new coding and noncoding genes and transposons among others [28]. Analyzed genomes were new *Oryza* species' chromosome-level reference assemblies, either long- or short-read technologies were derived from seven wild species (*O. barthii, O. glumaepatula, O. meridionalis, O. nivara, O. rufipogon, O. punctata, and Leersia perrieri*) and two domesticated varieties (*O. sativa* vg. *aus* [N22] and vg. *indica* [cv IR8] or miracle rice) in addition to previously published assemblies of *O. brachyantha, O. glaberrima, O. sativa* vg. *japonica* and *O. sativa* vg. *indica* [93–11]. Majority of the cultivars have AA genome, except for *O. punctata* (BB), *O. brachyantha* (FF), and *L. perrieri* (unknown), a closely related outgroup species.

The study identified a possible strong candidate and long-pursued Pi-ta2 locus [28] that can provide broad-specificity resistance to rice blast diseases upon interaction with Pi-ta (Bryan et al.,

2000). Among many blast disease pathogens, *Magnaporthe oryzae* pose major threats to global rice production of which yearly losses can potentially feed about 60 million people [28]. Overall, the accessibility of 13 high-quality reference genome assemblies will allow in-depth exploration of major orthologous loci and genomic regions of genotypes: AA, BB, FF, and *L. perrieri*.

5. The extent of rice genetic diversity

As previously discussed, several studies have shown the importance of germplasm resources in rice improvement and breeding programs for higher yields, better resistance to biotic and abiotic factors including the incidence of pests and diseases. Germplasm resources comprised of wild relatives, traditional, and modern cultivars among others are excellent sources of desirable traits in breeding programs. **Table 1** shows available information about useful and

Species	Genome	Characteristics/traits
O. barthii	AA	Resistance to BB, GLH; tolerance to heat and drought; drought avoidance
O. glaberrima	AA	Resistance to nematodes, rice yellow mottle virus, stem borers, African gall midge, iron toxicity, drought; tolerance to waterlogging; cultigen; crude protein content; weed competitiveness; high adaptability to acidic soils showing low levels of phosphorus availability
O. glumaepatula		Tolerance to heat; source of CMS; elongation ability
O. meridionalis		Tolerance to heat and drought; drought avoidance; elongation ability
O. nivara	AA	Resistance to BB, grassy stunt virus
O. rufipogon		Resistance to blast, Tungro virus, BB, BPH; moderately tolerant to Shb; increased elongation under deep water; Source of yield-enhancing loci and CMS; tolerance to aluminum and soil acidity
O. punctata	BB	Resistance to zigzag leaf hopper, BB, BPH; tolerance to heat and drought
O. brachyantha	FF	Resistance to YSB, BB, whorl maggot, leaf-folder; tolerance to laterite soil
Leersia perrieri	unknown	Stoloniferous; shade tolerance

Abbreviations: BB-bacterial blight; BPH-brown planthopper; CMS-cytoplasmic male sterility; GLH-green leafhopper; Shb-sheath blight; YSB-yellow stem borer.

Table 1. Important and useful genomic characteristics or traits in wild *Oryza* species with available reference genomes (sources: [33, 34]).

important traits in wild rice species. With reliable reference genome assemblies made available (like those studied in [28]), useful and important traits in wild *Oryza* species can be extensively used in wild hybridization and gene introgression toward breeding high-yield varieties expressing significant characteristics such as tolerance or resistance to biotic and abiotic stress factors that limit crop productivity in general.

In modern plant breeding, scientists repetitively rely on a continuous infusion of genetic resources to develop crops with superior resistance to diseases and pests, with better yield quantity and with the high quality of produce including flavor and nutritional values [10, 29–31].

Wild species	Key traits	Gene	Rice Variety
O. barthii	Drought-or heat-related traits	unknown	
O. glaberrima	Resistance to blast	unknown	Yun Dao, China-YAAS
	Tolerance to iron toxicity	unknown	many NERICA lines/varieties, Africa
	Drought-or heat-related traits	QTLs	
	Tolerance to acidic conditions	unknown	
	Tolerance to P deficiency	unknown	
	Tolerance to abiotic stresses, high yield, weed competitive ability, earliness		
O. glumaepatula	Cytoplasmic male sterility	unknown	
	Drought-or heat-related traits	unknown	
O. meridionalis	Drought-or heat-related traits	unknown	Arkansas rice varieties (2), USA
O. nivara	Resistance to grassy stunt	*GS*	many Asian rice varieties
	Resistance to bacterial blight	*Xa38*	
O. rufipogon	Resistance to bacterial blight	*Xa23*	Dhanarasi, India
	Resistance to blast	unknown	Matatag 9, Philippines
	Cytoplasmic male sterility	unknown	AS 996, Vietnam (acid sulfate)
	Tolerance to Tungro virus	unknown	BRRIdhan55, Bangladesh (salt)
	Tolerance to iron toxicity	unknown	
	Tolerance to aluminum toxicity	QTL	
	Tolerance to acidic conditions	unknown	
	Tolerance to P deficiency	unknown	
	Yield-enhancing loci	QTL, *yld1*, *yld2*	
	Increased elongation ability	unknown	
O. punctata	NO DATA		
O. brachyantha	Resistance to bacterial blight	unknown	
Leersia perrieri	NO DATA		

Table 2. Key traits of select wild *Oryza* species transferred in modern rice varieties through gene introgression and wide hybridization (source: [33]).

Conversely, scientists favor to employ existing cultivars or called elite or advanced breeding materials owing to previous experience in ease of intermating these cultivars and proven high productivity [31]. Aside from the difficulty of transferring outstanding traits, another major reason breeding scientists do not resort to using wild relatives in rice breeding is because of the inferior characteristics of wild relatives particularly, poor plant type and more grass-like appearance, poor grain type that shatters in nature, and low grain yield [31–33].

Nevertheless, successful breeding of wild rice species into modern rice varieties has been achieved through advances in tissue culture and molecular approaches [33] that complement conventional breeding methods. **Table 2** shows key traits from select wild rice species that have been transferred in modern rice varieties through gene introgression and wild hybridization.

Some examples of biotechniques used to explore favorable alleles in rice germplasm are employing advanced backcross populations to detect quantitative trait loci (QTL) associated with enhanced performance in rice as well as clone genes underlying key QTLs of interest (McCouch [30]) as well as use of backcross inbred lines (BILs) and chromosome segment substitution lines (CSSLs) and high-density single nucleotide polymorphism (SNP) arrays [32]. These research achievements further illustrate the importance of rice genetic resources in breeding new crops with highly desirable traits on top of high yield potential.

6. Molecular biotechniques to enhance rice breeding activities

Recent advances in molecular biology and biotechniques increase the chances of utilizing rice genetics resources that have not been explored in previous rice breeding programs. The availability of genomic, phenotypic, geographical, and ecological information among other sequence data, when analyzed all-together, can help researchers to strategically plan experiments based on developed models predicting plant performance [3].

Molecular approaches used in modern rice breeding include molecular marker technology and marker-assisted selection (MAS), molecular mapping of genes and QTLs and production of hybrids and alien introgression lines (AILs) [33] to name a few. One method of genomic-assisted breeding, MAS (**Figure 5**), utilizes molecular markers that map QTLs or specific genes known to be linked with phenotypes or target traits to choose individuals that exhibit desirable alleles for traits of interest [35]. Compared to conventional phenotypic selection, MAS has primary advantages, such as it is simpler than phenotypic screening, selection can be done at the seedling stage, and a single plant can be selected based on its genotype [36].

Another type of genomic-assisted breeding, called genomic selection (GS) utilizes all available marker data for a population as predictors of breeding value [35]. To generate a prediction model, GS combines marker data from a training population (TP) with phenotypic data as well as available pedigree data and then the model produces genomic estimated breeding values (GEBVs) of all TP-genotyped individuals [35]. GEBVs calculate the possible performance of a genotype as a potential parent in a breeding pipeline [35].

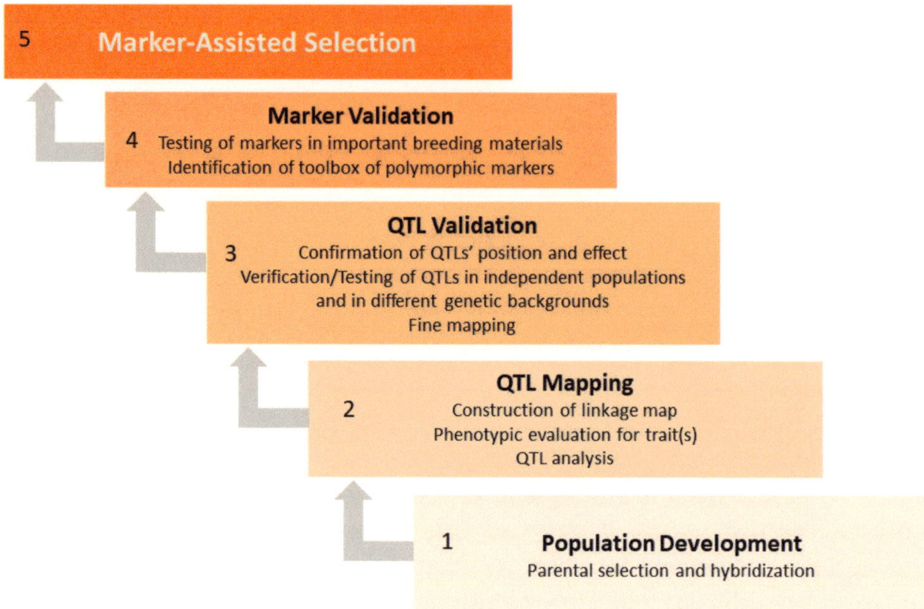

Figure 5. The process of marker development (adapted from [36]).

As modern plant biology further advances, other technology options like OMICs sciences or systems biology approaches become exploitable to pinpoint genomes (genomics) [37], genes (transcriptomics), metabolites (metabolomics), proteins (proteomics), and interactions (interactomics of protein-protein or protein-DNA) [38] among other complex biological systems that can revolutionize the improvement of crop productivity. For instance, metabolomics helps determine the differences between a healthy and a disease-infected plant through analysis of various levels of their thousands of molecules (ISAAA leaflet) [39]. Metabolomics can also identify plant defense metabolites and nutritional values as shown in a study that detected metabolites useful as molecular markers for drought stress tolerance for species related to tobacco and soybean [39].

In genomics research, present-day advances that made multi-layer genomic data derived from sequencing of both DNA and RNA now provide information on gene expressions, differential isoforms, alternative splicing methods, messenger RNA, non-coding RNA, and DNA polymorphism [37]. In short, current genomics is far ahead and progressive compared to traditional genomics that studies the function, structure, and sequence of a genome.

Recent studies in rice involving OMICs approaches have shown promise in scaling down the rice breeding process by focusing on the discovery of and engineering of desirable traits or genes. In China, scientists had integrated analyses of rice omics and biotechnological applications toward improving rice agronomic traits through molecular breeding approaches [40]. Integrated data analyzed include 220 functional genes that were cloned and identified, sequencing data of chromosome 4 of japonica rice Nipponbare and whole genome shotgun sequencing of Indica rice 9311 [40]. Additionally, the testing of comprehensive annotation platform for functional exploration of rice multi-omics data (CARMO) provided multiple web-based analysis tools for in-depth data mining and visualization [41]. CARMO stands for comprehensive annotation of rice multi-omics data. Its performance test showed useful functional insights for supplementary experimental studies and evidence that were previously reported [41].

A combinatorial analysis of data from genome-wide association study (GWAS) and high-throughput phenotyping was used to evaluate the effects of salinity in rice particularly on relative growth rate (RGR), transpiration rate (TR), and transpiration use efficiency (TUE) [42]. Results include the identification of new candidate genes responsible for the early response of rice to salinity stress (**Figure 6**). Through interaction model approach, early response of rice to salt stress was associated with various signaling mechanisms as shown by expressions of signaling-related genes such as *Os03g16130* (encoding a calcium/calmodulin *(Ca/CaM)-dependent kinase*), *Os05g39870* (encoding *OsCIPK28* and *CAMK_KIN1, CA/CaM-dependent protein kinase*), *Os05g39900* (encoding a CBL-interacting serine/threonine-protein kinase 15), *Os05g46320* (encoding *OsFBX173*, an F-box domain-containing protein), and *Os05g47670* (containing a

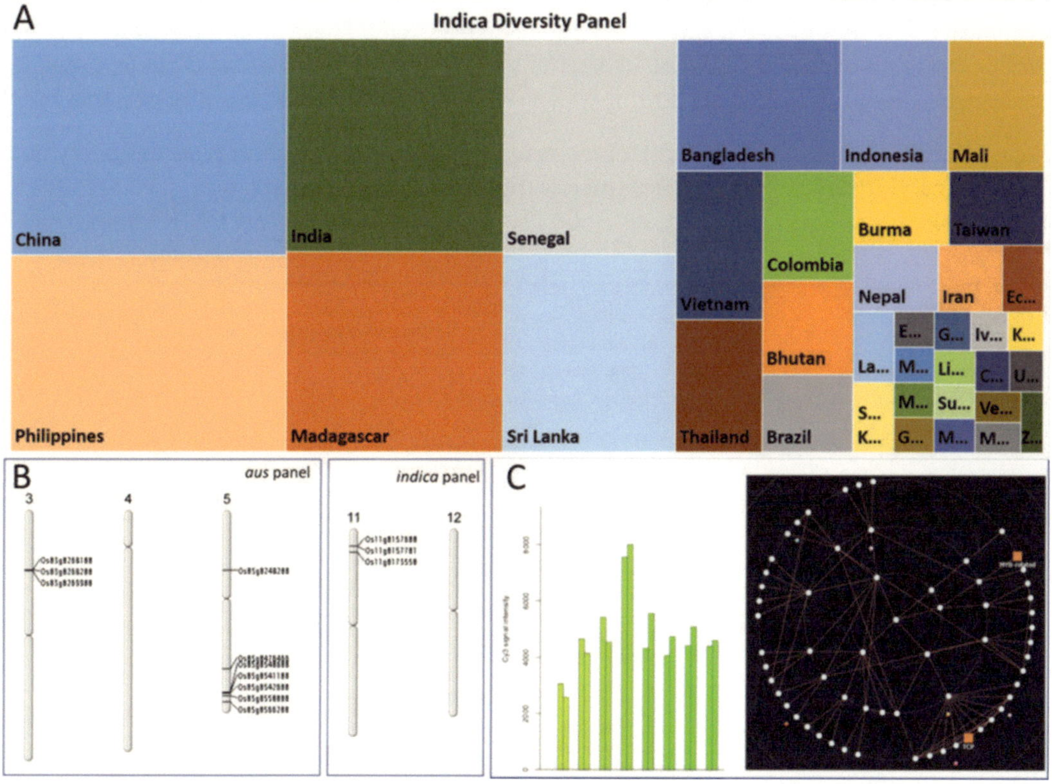

Figure 6. Utilization of rice diversity panel for identification of salinity tolerance loci in the study of Al-Tamini et al. [42]. (A) Blocks represent source countries of germplasm sizes contributed in the diversity panel (data taken from GRiSP global Rice Phenotyping network). (B) Candidate genes involved in early salinity response in aus and indicator panels. (C) Expression profile of *Os03g16130* (*calcium/Calmodulin-dependent kinase*) in rice inflorescence and co-expression network (expression and network data sourced from RiceXPro [44] and RiceFREND [45]).

zinc-finger motif, a C3HC4-type domain-containing protein) [42]. Candidate genes and QTLs can also be detected effectively using GWAS and gene-based association analysis followed by haplotype analysis for forthcoming functional characterization and genetic improvement of protein content, consumption, and cooking quality [43].

7. Rice genetic resources: experiences and prospects

Overall, rice genetic resources have been used to improve modern cultivars, although common knowledge is 95% of the rice genepool [27] remains untapped and unexploited in rice improvement. Major reasons of breeders for not using a wider genepool base including wild relatives are inferior traits of wild relatives in terms plant type, grain shattering, and low yield and generally, the difficulty of transferring desirable traits as well [31–33].

As more information about germplasms is available, wider selection and diversity of materials can be exploited for varietal improvement [46]. Generating more usable information becomes more possible with modern biotechniques that can examine and identify specific genes, genomes, QTLs, and proteins among other genetic materials. These techniques target precise genetic materials expressing advantageous traits toward increasing yield and quality, specific resistance to biotic and abiotic factors as well as emerging or future production constraints.

Figure 7 shows a concept framework of molecular- and systems-levels of gene discovery and transfer enabling development of new breeds from a wide genepool including wild relatives. Various biotechnologies such as GWAS, QTLs, genomic selection, and OMICs approaches

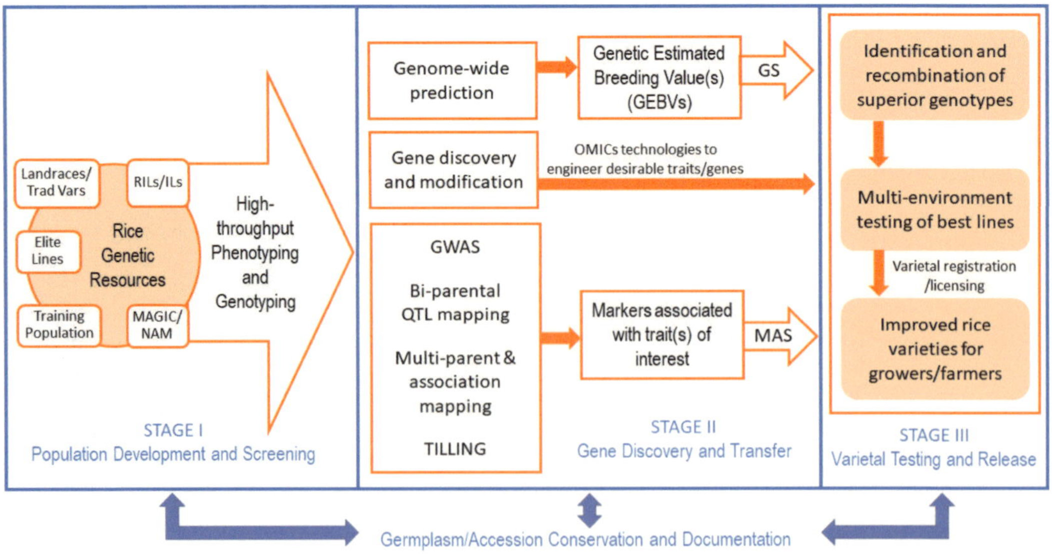

Figure 7. The conceptual framework of the applications of NGS and OMICs technologies for molecular- and systems-level breeding of improved crop varieties. Seeds and information generated from stages 1 to 3 generation and submission of information to genebank for the provision of better passport and characterization data of accessions or germplasm for future breeding use.

among other techniques complementing conventional breeding methods can hasten the discovery and transfer of desired genetic materials toward developing improved varieties.

Compared to yesteryears, these biotechniques are becoming low-cost and easier to use and can now be routinely employed to identify and transfer desirable traits into new crops. In a recent review about genetically engineering crops, OMICs approaches were evaluated to weigh in the intended and unintended effects of genetic engineering (GE) in plant breeding. Results suggest that GE and OMICs technologies have remarkable potentials toward boosting crop improvement initiatives in the twenty-first century in conjunction with conventional breeding techniques [47].

Figure 7 also points out the importance of conserving seeds as germplasm and providing information derived from biotechniques for further use and reference in future breeding programs. For example, genotypic and phenotypic information will further elucidate the values of germplasm resources [6]. Accessibility of integrated germplasm characterization information including molecular markers, genome sequence information, genotype-phenotype relationships will entice breeders to use a wider genepool as potential parent lines and as sources of important traits and characteristics. Development of integrated germplasm information for a germplasm should be a collaborative effort of scientists, such as physiologists, geneticists, pathologists, genebank curators, and breeders, who share parallel and harmonized interests in exploiting the genetic materials of germplasm resources and determining valuable and novel genes for crop improvement.

In summary, advances in complementary plant breeding methods including molecular marker technology, OMICs approaches have allowed rapid developments in basic knowledge of genetics and breeding of plants. Molecular approaches have also provided important information about genetic materials and specific genes and genomes for the development of improved crops. Additionally, these approaches hold promise and help speculate about how their applications can shape the future of rice breeding for beneficial prospects of rice producers and consumers worldwide. The ease of use of these biotechniques coupled with affordability will allow better characterization of genetic resources that can increase their utility to improve crops to address present-day and future production constraints.

Author details

Roel C. Rabara[1*], Marilyn C. Ferrer[2], Jennifer Jara-Rabara[2] and Rachel C. Sotto[3]

*Address all correspondence to: roel.rabara@jacks.sdstate.edu

1 New Mexico Consortium, Los Alamos, USA

2 Philippine Rice Research Institute, Science City of Muñoz, Maligaya, Nueva Ecija, Philippines

3 Institute of Biological Sciences, University of the Philippines Los Baños, Laguna, Philippines

References

[1] Frison C, López F, Esquinas-Alcazar J. Plant Genetic Resources and Food Security: Stakeholder Perspectives on the International Treaty on Plant Genetic Resources for Food and Agriculture Organization (FAO) of the United Nations. New York, NY. Bioversity International and Earthscan. Routledge; 2012

[2] Food, C.o.G.R.f. The Second Report on the State of the world's Plant Genetic Resources for Food and Agriculture. Food & Agriculture Organisation (FAO). Rome, Italy. 2010

[3] McCouch S et al. Agriculture: Feeding the future. Nature. 2013;**499**(7456):23-24

[4] Platform, C.G., Genebanks Platform: Full Proposal 2017-2022 Addenda; 2016

[5] Galluzzi G et al. Twenty-five years of international exchanges of plant genetic resources facilitated by the CGIAR genebanks: A case study on global interdependence. Biodiversity and Conservation. 2016;**25**(8):1421-1446

[6] McCouch SR et al. Genomics of gene banks: A case study in rice. American Journal of Botany. 2012;**99**(2):407-423

[7] Plucknett D et al. Gene Banks and the world's Food. Gene Banks and the world's Food. Princeton, NJ: Princeton Univ. Press; 1987

[8] Rabara R et al. Conservation of rice genetic resources for food security. Advances in Food Technology and Nutritional Sciences. 2015;SE(1):S51-S56. DOI: 10.17140/AFTNSOJSE-1-108

[9] Maxted N et al. Towards a definition of a crop wild relative. Biodiversity and Conservation. 2006;**15**(8):2673-2685

[10] Guarino L, Lobell DB. A walk on the wild side. Nature Climate Change. 2011;**1**(8):374

[11] Vollbrecht E, Sigmon B. Amazing grass: Developmental genetics of maize domestication. Biochemical Society Transactions. 2005;**33**(Pt 6):1502-1506

[12] Vincent H et al. A prioritized crop wild relative inventory to help underpin global food security. Biological Conservation. 2013;**167**:265-275

[13] IRRI. The International Rice Genebank. 2018 [cited 2018 February 27]; Available from: http://irri.org/our-work/research/genetic-diversity/international-rice-genebank

[14] Khoury CK et al. Estimation of countries' interdependence in plant genetic resources provisioning national food supplies and production systems. The International Treaty on Plant Genetic Resources for Food and Agriculture. Rome, Italy. 2015

[15] Conway G. The Doubly Green Revolution: Food for all in the twenty-First Century. Ithaca, New York, USA: Cornell University Press; 1998

[16] Fowler C, Hodgkin T. Plant genetic resources for food and agriculture: Assessing global availability. Annual Review of Environment and Resources. 2004;**29**(1):143-179

[17] Awika JM. Major Cereal Grains Production and Use around the World, in Advances in Cereal Science: Implications to Food Processing and Health Promotion. Washington, DC, USA: ACS Publications; 2011. pp. 1-13

[18] Childs N. Rice Outlook. Washington DC: USDA-ERS; 2017

[19] OECD/FAO. Agricultural Outlook 2015. Paris: OECD Publishing. p. 2015

[20] John A, Fielding M. Rice production constraints and 'new' challenges for south Asian smallholders: Insights into de facto research priorities. Agriculture & Food Security. 2014;**3**(1):18

[21] Crowley E. Ending Poverty: Learning from Good Practices of Small and Marginal Farmers. Self Employed Women's Association Exposure and Dialogue and the Food and Agricultural Organization. Rome, Italy. 2013

[22] Waddington SR et al. Getting the focus right: Production constraints for six major food crops in Asian and African farming systems. Food Security. 2010;**2**(1):27-48

[23] Chauhan BS, Johnson DE. Row spacing and weed control timing affect yield of aerobic rice. Field Crops Research. 2011;**121**(2):226-231

[24] Alibu S et al. Farmer's knowledge and perceptions on Rice insect pests and their Management in Uganda. Agriculture. 2016;**6**(3):38

[25] Marshall E et al. Climate Change, Water Scarcity, and Adaptation in the US Field Crop Sector. United States Department of Agriculture, Economic Research Service. USDA-ERS, Washington, DC, USA. 2015

[26] De Leon J, Carpena A. Pedigree-based genetic diversity analysis of improved rice (*Oryza sativa* L.) varieties in the Philippines. Philippine Journal of Crop Science (Philippines). 1995

[27] Xu W et al. Molecular diversity and its geographical distribution in core rice germplasm. In: Advances in Rice Genetics: (In 2 Parts). World Scientific; 2003. pp. 182-184

[28] Stein JC et al. Genomes of 13 domesticated and wild rice relatives highlight genetic conservation, turnover and innovation across the genus Oryza. Nature Genetics. 2018:1

[29] Castañeda-Álvarez NP et al. Global conservation priorities for crop wild relatives. Nature Plants. 2016;**2**(4):16022

[30] McCouch SR et al. Through the genetic bottleneck: O. Rufipogon as a source of trait-enhancing alleles for O. Sativa. Euphytica. 2007;**154**(3):317-339

[31] Rubenstein KD, Heisey P. Plant Genetic Resources: New Rules for International Exchange. 2003 [cited 2018 February 27]

[32] Ali ML et al. Chromosome segment substitution lines: A powerful tool for the introgression of valuable genes from Oryza wild species into cultivated Rice (O. Sativa). Rice. 2010;**3**(4):218-234

[33] Sanchez PL, Wing RA, Brar DS. The wild relative of rice: Genomes and genomics. In: Zhang Q, Wing RA, editors. Genetics and Genomics of Rice. New York, NY: Springer New York; 2013. pp. 9-25

[34] Wambugu PW et al. Conservation and utilization of African Oryza genetic resources. Rice. 2013;**6**(1):29

[35] Varshney RK, Terauchi R, McCouch SR. Harvesting the promising fruits of genomics: Applying genome sequencing technologies to crop breeding. PLoS Biology. 2014;**12**(6):e1001883

[36] Collard BCY, Mackill DJ. Marker-assisted selection: An approach for precision plant breeding in the twenty-first century. Philosophical Transactions of the Royal Society, B: Biological Sciences. 2008;**363**(1491):557-572

[37] Yang MQ et al. The emerging genomics and systems biology research lead to systems genomics studies. BMC Genomics. 2014;**15**(11):I1

[38] Tripathi P, Rabara RC, Rushton PJ. Emergence of Omics-approaches for crop improvement during abiotic stress. International journal of Environmental Science and Technology. 2012;**1**(2):125-128

[39] Rabara RC, Tripathi P, Rushton PJ. Comparative Metabolome profile between tobacco and soybean grown under water-stressed conditions. BioMed Research International. 2017;**2017**:12

[40] Li J et al. Rice omics and biotechnology in China. Plant Omics. 2011;**4**(6):302

[41] Wang J et al. CARMO: A comprehensive annotation platform for functional exploration of rice multi-omics data. The Plant Journal. 2015;**83**(2):359-374

[42] Al-Tamimi N et al. Salinity tolerance loci revealed in rice using high-throughput non-invasive phenotyping. Nature Communications. 2016;**7**

[43] Wang X et al. Genome-wide and gene-based association mapping for rice eating and cooking characteristics and protein content. Scientific Reports. 2017;**7**(1):17203

[44] Sato Y et al. RiceXPro version 3.0: Expanding the informatics resource for rice transcriptome. Nucleic Acids Research. 2012;**41**(D1):D1206-D1213

[45] Sato Y et al. RiceFREND: A platform for retrieving coexpressed gene networks in rice. Nucleic Acids Research. 2012;**41**(D1):D1214-D1221

[46] Romero GO et al. Development and utilization of the PhilRice Genebank. Philippine Science Letter. 2011;**4**(1):24-32

[47] National Academies of Sciences, Engineering and Medicine. Genetically Engineered Crops: Experiences and Prospects. Washington, DC, USA: National Academies Press; 2016

Section 3

Nanotechnology in Agriculture

Chapter 6

Detection Methods of Nanoparticles in Plant Tissues

An Yan and Zhong Chen

Additional information is available at the end of the chapter

http://dx.doi.org/10.5772/intechopen.74101

Abstract

The increasing use of nanoparticles (NPs) in the world has raised significant concerns about their potential impacts on ecosystems, food safety and human health, leading to an emerging research theme about the interaction between crop plants and NPs. Therefore, a full understanding of plant-NP interaction and phytotoxicological mechanism is required for accurate risk assessment to ensure the safe use of nanoparticle. A range of analytical techniques have been developed to detect and characterize the uptake, translocation, cellular internalization and intracellular biotransformation of nanoparticles in plants. Imaging methodologies, including various electron microscopy, spectrometry-based techniques, together with ICP-based techniques such as ICP-OES, ICP-MS and SP-ICP-MS, have been widely used to obtain information about NPs size, morphology, size distribution, cellular localization, elemental speciation, mass concentration and so on. Due to the complexity of biological samples to be analyzed, these techniques are often combined accordingly to provide complementary information regarding plant-NP interaction. This review provides an introduction to the most widely used techniques in the study of interactions between plants and nanoparticles. In addition, applications of these techniques in the study of plant-NP interaction from recent works are exemplified to illustrate how the understanding of plant-NP interaction is achieved through these techniques.

Keywords: nanoparticles, microscopy, X-ray, mass spectrometry, plant, uptake, biotransformation

1. Introduction

Over the past decades, nanotechnology has been widely applied on commercial products on the market, including biosensor, catalysts to optics, antimicrobial activity, computer transistors, electrometers, and wireless electronic logic and memory schemes. In the agriculture sector, nanoparticles are often incorporated into nano-formulated pesticides, fertilizers, and nanobiosensors for crop protection [1]. Therefore, the application of engineered nanomaterials worldwide inevitably caused the release and accumulation of nanoparticles in the environment [2].

As the basic components of the ecosystem, plants are sessile and their roots absorb nutrients and water as well as contaminants from their environment. Accumulating evidence demonstrated that engineered nanoparticles could be released from some commercial products, further be taken in and accumulated in plant tissues. As plants may serve as a potential pathway for the transportation of nanoparticles through the food chain [3], the increasing applications of engineered nanomaterials in the world have raised a growing concern about their potential adverse impacts on ecosystems, food safety and human health [2, 4]. Therefore, to evaluate potential environmental risks imposed by nanoparticle application, it is important to understand the interaction between nanoparticles and plants, as well as NP's behavior and toxicity in plants. However, the behavior of NPs in plants and phytotoxicity mechanism are so complicated that contradictory results regarding the effects of nanoparticles on plants were obtained from various studies during the past decade [5–9]. These conflicting results indicate that impacts of nanoparticles on plants largely depend on the type and concentration of nanoparticles, plants species, tissue exposed, and the experimental conditions [6, 10, 11].

There are various engineered nanoparticles with different size, morphology, and properties. Engineered nanoparticles also exhibit distinct physical and chemical properties with different environmental behaviors and toxicity in comparison with their bulk counterparts, which could be attributed to the small size at nanoscale (1–100 nm) and high surface-to-volume ratios of nanoparticles [12]. Upon nanoparticle exposure, the directly contact between nanoparticles and roots leads to the uptake of nanoparticles by roots and translocation of nanoparticles in plants [10, 13]. Different types of nanoparticles exhibit distinct behaviors and translocation characteristics. During interaction with biological environments, nanoparticles can also be transformed by plants, which in turn alter environmental fate and toxic properties of nanoparticles [14, 15]. Therefore, the toxic effect and behavior of nanoparticles are determined by not only the initial properties (such as particle size, shape, structure, charge, elemental composition, mass concentration, and state of aggregation etc.), but also by the physicochemical evolution [16–18]. Hence, in order to accurately assess the phytotoxicity of nanoparticles, it is necessary to determine the original characteristics of NPs before treatment, uptake and translocation, cellular internalization and intracellular biotransformation during interaction with plants.

Approaches to detect and characterize NPs during plant-NP interaction are thus becoming crucial in our studies. Nowadays, a variety of analytical techniques have been developed to provide the necessary information regarding plant-NP interaction, including microscopy imaging, chromatography, spectrometry-based techniques, and so on. In this review, we describe the advantages and limitations of a selection of current most frequently-used methods in the study of uptake, distribution, translocation and biotransformation of NPs in plants. We also exemplify the usage of these analytical techniques with instances from recent studies.

2. Techniques for nanoparticle detection

2.1. Imaging techniques

2.1.1. Transmission electron microscopy/scanning electron microscopy

Transmission electron microscopy (TEM) and scanning electron microscopy (SEM) are considered to be the most popular techniques for the analysis of nanomaterials. Electron microscopy straightforwardly captures projected area of the particles, providing visualization of true particle size dimensions.

In TEM process, a focused electron beam is transmitted through a specimen, an image is formed from the interaction of the electrons with the sample. The image is then magnified and focused onto an imaging device. TEM is capable of imaging at a significantly higher resolution (down to the sub-nanometer) than light microscopes, it can visualize as small as a single column of atoms, which is thousands of times smaller than a resolvable object seen in a light microscope.

In SEM process, a focused beam of electrons scans the surface of the sample; interaction between electrons and atoms in the surface of sample produces various signals that contain information about the sample's surface topography and composition. Then an image is formed upon focusing of scattered electrons. SEM can achieve resolution better than 1 nm [19]. It can be used to image intact sample as well as sectioned sample [15].

Through visualization of nanoparticle position within a cell or tissue, TEM/SEM can provide the precise nanoparticle information about their size, structure, shape, morphology, dispersion or aggregation state, which is informative for assessing in vitro nanoparticle uptake and localization. Nanoparticle sizes are calculated and expressed as a sphere diameter having a similar projected area as the projected image of the nanoparticle. Particle size analysis is carried out by manually using a marking device to move along the nanoparticles. A mean linear dimensional measure of the nanoparticles is obtained by dividing the total length of the nanoparticles by the total number of nanoparticles counted [20]. In addition, when combined with spectroscopic methods, characterization of the composition of the internalized nanoparticles became possible [21]. Owing to the high lateral resolution of TEM, it could also be used to trace the dynamics of individual NPs in a living cell or plant tissues.

2.1.2. Scanning transmission electron microscope

Scanning transmission electron microscope (STEM) is a type of transmission electron microscope (TEM). A typical STEM is a conventional TEM equipped with additional scanning coils, detectors and necessary circuitry. Like a conventional TEM, images are formed by electrons transmitting through a thin specimen. The difference is that in STEM the electron beam is focused to a fine spot (with spot size 0.05–0.2 nm), then it scans over the sample in a raster. The rastering of the beam across the sample makes STEM suitable for combination with analytical techniques such as annular dark-field imaging and spectroscopic mapping, to obtain information on the structure of nanoparticles with sub-nanometer resolution and their chemical composition [22, 23].

Dark-field microscopy with a STEM, such as high-angle annular dark field scanning transmission electron microscopy (HAADF-STEM), can be used to distinguish elements with high atomic number (Ag, Au, etc.) from the major elements in organisms (C, N, O, etc.) with a high spatial resolution (down to 1 nm) [24].

2.1.3. Dynamic light scattering

Dynamic light scattering (DLS) technique is the most commonly employed high-throughput technique to measure nanoparticle size and determine aggregation state of nanoparticles in aqueous suspensions. In DLS analysis, the Brownian movements of the NPs in aqueous suspensions cause constructive and destructive interference, which results in time-dependent fluctuations in scattering intensity. Then the average particle size can be calculated from these time-dependent fluctuations in scattering intensity by application of the autocorrelation function and subsequent calculation of the exponential decay [25]. Meanwhile, the zeta potential, a key indicator of the stability of colloidal dispersions, is measured rapidly using DLS. DLS is able to analyze samples containing very broad distributions of species; it can also detect very small amounts of the higher mass species [25].

2.1.4. Energy-dispersive X-ray spectroscopy

Energy-dispersive X-ray spectroscopy (EDS) is used as an analytical technique to analyze a sample's elemental composition or chemical characterization. EDS applies a high-energy electron beam which focuses into the sample of interest to excite in an inner shell an electron and to eject it from the shell, thus generating an electron hole. Then an electron from the outer shell with higher-energy fills the hole, which releases the energy in difference in between the outer shell and the inner shell in the form of X-ray [26]. As each element has a unique set of peaks on electromagnetic emission spectrum determined by its unique atomic structure, through measuring the number and energy of the X-rays emitted from the sample by an EDS instrument, the information about elemental composition of the sample is obtained [26].

2.1.5. X-ray absorption and X-ray fluorescence

There are two main types of X-ray spectroscopy-based techniques that can be used to analyze speciation and localization of NPs within the plant tissue: X-ray fluorescence (XRF) spectrometry and X-ray absorption (XAS) spectrometry. Both of them are based on measuring the spectra of emission or absorption of X-radiation. The absorption of X-ray photons by element is controlled by the photo-electric effect.

When sample is subjected to X-ray radiation, incident X-rays (photons) of a definite energy shine on the samples. If the energy of incident X-rays that reaches the sample is lower than the binding energy (E0) of the core electrons of the element, the atoms of this element do not participate in the absorption process. While with increasing energy of the incident X-ray photons, a point will be reached where their energy is approximately equal to the binding energy of the core electrons. At this point a sharp increase in absorption of the X-ray photons occurs [27]. The energy absorbed by the core electron elevates it into a higher energy state or electron orbital, which is unoccupied. This excited core electron is referred to as a photoelectron. At the binding energy (E0), the photoelectron is ejected from the atom into the continuum [28]. As a result, a vacancy in the shell of the core electron is created in a core orbital. In order for the atom to return to the ground state, an electron from a higher energy orbital (e.g. L or M) fills the vacancy consequently, emitting X-ray photons in the form of fluorescence with characteristic energy corresponding to the difference between the two electronic levels' binding energies. These X-ray photons have characteristic energies for each element in the periodic table, confers element-specificity to the absorption and fluorescence spectroscopy.

During XRF process, the emitted fluorescence signal can be recorded at each position and used to generate XRF elemental maps. XRF is a nondestructive technique, it can be used to identify and determine the concentrations of elements present in biological samples, as well as providing information of in situ localization of elements in the samples.

During XAS process, the energy of the incident X-ray beam is progressively increased beyond the binding energy, thus the emission of fluorescence and absorption of the incident X-ray progressively increases, generating a characteristic X-ray absorption spectrum by detecting and recording the absorption or fluorescence at each energy point. The main feature of the XAS spectrum is a sharp, step-like curve called the absorption edge [28, 29].

The XAS spectrum is conventionally divided into two parts according to the energy region. The region comprising the pre-edge, the edge-jump and post-edge covering the energy range from approximately −50 to +50 eV of the absorption edge is defined as X-ray absorption near edge structure (XANES). The region from +50 to +1000 eV above the edge is defined as extended X-ray absorption fine structure (EXAFS) [28]. XANES is particularly sensitive to the oxidation states of elements and the electronegativity of the ligands, and it provides electronic structural information about the oxidation state and local geometry of the absorbing metal atom. EXAFS can provide information about the element coordination such as the identity and number of the coordinating atoms, and the interatomic distance between the central absorbing atom and its next nearest neighbors.

XAS is an element specific spectroscopic technique that provides specific qualitative information about chemical species at very high (subatomic) spatial resolution and is able to analyze almost any type of samples including amorphous (non-crystalline) materials in situ, requiring minor or no sample preparation prone to modify the chemical species. XAS experiments require an intense and polychromatic X-ray source. Synchrotron radiation is a very intense, collimated and polarized X-ray source, with a continuous band of wavelengths from around the μm (infrared) to the pm (hard X-ray) range [30]. Nowadays, owing to the development of synchrotron radiation facilities, the combination of synchrotron radiation with XAS is proved to be a powerful technique for speciation analysis of chemical elements.

XAS spectroscopy can be performed as bulk analyses to assess the overall speciation of the chemical of interest in the sample (usually homogenized). For bulk-XAS experiment, a beam in the size range of a few hundred μm^2 to a few mm^2 is used to illuminate sample, the XAS spectra obtained are generally representative of the average speciation of the chemical in the sample [28, 31]. Whereas, the signal of minor species in the sample, which accounts for less than 5–10% of the total analyte, is insufficient to be resolved and quantified from the bulk spectra. In this case, bulk-XAS analysis is insufficient for us to obtain specific information from a complex and heterogeneous mixture of biological sample. This limitation can be overcome by decreasing the beam size to the range of tens of nm to a few um, and using thin section of sample. This kind of laterally-resolved XAS analysis is referred to as μ-XAS.

From μ-XAS spectra, the information attained at each point of analysis is only representative of the spot probed and not of the overall speciation in the sample. Therefore, a trade-off exists between detecting minor species and obtaining the overall speciation of the analyte in the whole sample. A strategy to solve the problem is coupling bulk XAS analyses with laterally resolved techniques such as μ-XRF, μ-XAS and μ-XRD. In a typical work flow, laterally resolved μ-XRF elemental maps are first collected to identify spots of interest, which are then further probed by μ-XAS analysis [28].

Recently, a new approach termed XANES imaging has been developed with the capacity to analyze element speciation and full lateral distribution over large areas of sample. In a XANES imaging process, an elemental map of the sample will be firstly generated to identify interesting areas, then the μ-X-ray fluorescence signals from the interesting areas are collected repeatedly over progressively increased incident X-ray energies and scan across the characteristic absorption edge of the target element. These resulting maps can be aligned and stacked, then the XANES spectra can be extracted from individual pixels or groups of pixels over regions of interest, eventually, the spatial distribution of both major and minor species within the sample will be obtained [28]. A detailed information regarding comparison among bulk-XANES, μ-XANES and XANES imaging is provided in a review by Gräfe et al. [28].

2.1.6. X-ray diffraction

X-ray diffraction (XRD) is a nondestructive technique for characterizing crystallographic structure or elemental composition of crystalline materials. It can reveal information about the crystal structure, crystalline phase, preferred crystal orientation (texture), average crystallite size and strain of materials. The constructive interference of a monochromatic beam of X-rays diffracted at specific angles from each set of lattice planes in the crystalline sample will produce X-ray diffraction peaks, intensities of which are determined by the distribution of atoms within the lattice, therefore, an X-ray diffraction pattern will be generated which reflects the periodic atomic arrangements in the sample.

For synchrotron-based X-ray diffraction (SR-XRD) technique, the high intensity and well-defined wavelength of the incident synchrotron radiation will generate a better resolution of diffraction peaks and make SR-XRD capable in detecting minor constituents in a sample [27]. In addition, XRD is capable of 20 μm lateral resolution with minimal sample preparation requirements, can be used as a valuable complementary or alternative methods to XAS analysis. A limitation of this method is that it's not applicable for amorphous materials; it can only characterize crystalline samples.

2.1.7. X-ray computed microtomography

X-ray computed microtomography (μCT) uses X-ray to create cross-sections of a sample that can be used to produce three-dimensional digital images of the sample's internal structure at a micron level spatial resolution without destroying the original sample [32].

In an absorption-edge synchrotron radiation-based μCT process, a high flux, monochromatic X-ray beam passes through the sample, a scintillator converts the transmitted X-rays into visible light and the resulting absorption projection is captured by a photodetector to produce 2D radiographs. The sample is then rotated (or the X-ray source and detector are rotated about the object) by a small angle, a series of 2D X-ray absorption images is captured successively between 0° and 180°. Using mathematical principles of tomography, this series of images is then reconstructed to produce a 3D image, thus a 3D distribution of the element of interest within the sample is obtained [27, 32].

2.1.8. Scanning transmission X-ray microscopy

Scanning transmission X-ray microscopy (STXM) is a type of X-ray microscopy that allows in situ mapping of elements at high lateral resolution within a specimen. STXM uses a Fresnel zone plate to focuses synchrotron soft X-ray absorption beamline into a small spot, the sample is placed at the focus of the zone plate and scanned by X-ray, then a film or charged coupled device detector is used to detecting the transmitted X-rays intensity that pass through the specimen [33].

STXM-XAS, a technique that in-situ conditions of a XAS experiment with a STXM microscope, is capable of determining chemical speciation with a spatial resolution of 10–30 nm [34]. STXM-XAS can handle samples with thicknesses up to 20 micron at 1.5 keV, which makes it possible to study a wider and more flexible range of materials, including various plant tissues [33].

2.1.9. Nano secondary ion mass spectrometry

Secondary ion mass spectrometry (SIMS) uses an energetic ion beam to bombard a sample, particles from the top few atomic layers of the sample surface are then removed, resulting in the consequent liberation of ions, known as secondary ions. These secondary ions are then sorted on the basis of their energy in the instrument's electrostatic sector and later dispersed in a mass spectrometer to produce a map giving information about the elemental or molecular distribution within the sample [29, 35].

Nano secondary ion mass spectrometry (NanoSIMS) is a nanoscopic scale resolution chemical imaging mass spectrometer based on SIMS [35]. The main advantage of NanoSIMS over other SIMS is the ability to operate at high mass resolution, while maintaining both excellent signal transmission and high lateral resolution (down to 50 nm) with a low detection limit (mg/kg range). It is capable of measuring most elements in the periodic table, from hydrogen to uranium, as well as their different isotopes. These advantages of NanoSIMS make it one of the most powerful tools to quantitatively investigate elemental distribution in organisms at the cellular level [36, 37]. It is reported that Nano-SIMS has been used for the analyses of NPs in biological samples including plant tissue [36].

Technique	Information provided	Spatial resolution	Detection limit	Advantages	Limitations	Examples of application
TEM	Size, size distribution, shape, distribution, aggregation state, structure	>0.1 nm	mg/kg	High resolution, in vivo	Destructive, sample preparation, high vacuum condition, insensitive to light elements	Cucumber [54]
SEM	Size, size distribution, shape, distribution, aggregation state, structure	1 nm to 1 μm	mg/kg	High resolution, in vivo	High vacuum, sample preparation, insensitive to light elements	*Eichhornia crassipes* [50]
STEM	Size, size distribution, shape, distribution, aggregation state, structure	<0.1 nm	mg/kg	Atomic resolution, analysis of low concentrations (ppm), in vivo	Sample preparation, insensitive to light elements	*Chlamydomonas reinhardtii* [36]
DLS	Size distribution, zeta potential, hydrodynamic diameter	3 nm-μm		In situ and real-time measurement, rapid and simple analysis	Difficult to interpret heterogeneous size distributions, aggregates, dust particles can ruin the measurements on nanoparticles, multiple scattering and particle interactions in high concentrations, limited capability on polydisperse samples	*Capsicum annuum* L. [47] Romaine lettuce [48]
EDS	Elemental composition, distribution			Nonquantitative analysis		
XAS	Oxidation state, elemental composition, structure	ppm	mg/kg	In vivo, minor species can be investigated	Cause beam damage artifacts	Cucumber [54]
XRF	Solid state speciation, quantitative bulk analysis, isotope ratios, morphology		mg/kg	Nondestructive, in vivo		*Landoltia punctata* [51]
XRD	Structure, size	1–3 wt%		Nondestructive		
μ CT	Distribution			In vivo, nondestructive, 3D visualization		Wheat [52]

http://dx.doi.org/10.5772/intechopen.74101

Technique	Information provided	Spatial resolution	Detection limit	Advantages	Limitations	Examples of application
STXM	Size, shape, visualization	30 nm		In vivo, no sample preparation, nondestructive, liquid conditions		Cucumber [59]
NanoSIMS	Distribution, elemental composition, surface properties	50 nm		In vivo, high mass resolution, high lateral resolution	Destructive, difficult to analyze some elements with poor secondary ion yield, such as Zn, Cd, and Mn	*Chlamydomonas reinhardtii* [36]
ICP-OES	Elemental composition, concentration	Not available	μg/l	Quantitative analysis	Do not provide information on particle shape or diameter	Rice [53]
ICP-MS	Bulk elemental composition, number concentration, mass concentration	Not available	ng/l	Quantitative analysis, high sensitivity, low background signal, rapid and simple analysis	Do not provide information on particle shape or diameter, minimum particle size is limited by ICP-MS sensitivity, background and dissolved element content, narrow optimum range of particle number concentrations	*Arabidopsis thaliana* [49]
SP-ICP-MS	Concentration, number concentration, mass concentration	Not available	ng/l	Quantitative analysis, high sensitivity, low background signal, rapid and simple analysis	Do not provide information on particle shape or diameter, minimum particle size is limited by ICP-MS sensitivity, background and dissolved element content, narrow optimum range of particle number concentrations	Soybean and rice [55] *Arabidopsis thaliana* [44]

Table 1. Overview of analytical techniques discussed in this review with examples of application.

A limitation of NanoSIMS technique is that it is difficult to analyze elements with poor secondary ion yield, such as Zn, Cd, and Mn [37]. In addition, it is a destructive technique, which can be a disadvantage for some samples. This problem can be overcome by using high-pressure freezing followed by freeze substitution to preserve cellular and subcellular structures as well as elemental distributions of plant cells [29].

2.2. ICP-based techniques

2.2.1. Inductively coupled plasma-optical emission spectrometry

Inductively coupled plasma (ICP) based analytical techniques can provide quantitative elemental composition of a wide variety of sample types, including solids, liquids, and suspensions. Inductively coupled plasma-optical emission Spectrometry (ICP-OES) can be used to measure nanoparticle number concentration and elemental composition within a sample. As ICP-based techniques involve the use of liquid phases, suspensions could be analyzed directly, but solid samples have to be pretreated for the digestion of the matrix [21]. Generally, solid samples are dissolved or digested using acid in a microwave to get volatile analytic species. The sample solution is then nebulized into the core of inductively coupled argon plasma, where a flame temperature in a range from 6000 to 10,000 K vaporizes the nebulized solution, thus the analytic species are atomized, ionized and thermally excited. The excited atoms and ions return to low energy position, emitting electromagnetic radiation at wavelengths characteristic of a particular element, then the analytic species can be detected and quantified with an optical emission spectrometer (OES) through measuring the intensity of radiation, which is converted to elemental concentration by comparison with calibration standards.

2.2.2. Inductively coupled plasma-mass spectrometry

Inductively coupled plasma-mass spectrometry (ICP-MS) is an inorganic elemental analysis technique based on atomic mass spectrometry. ICP-MS consists of an ion source, a sampling interface, ion lens, a mass spectrophotometer and a detector system [18]. ICP sources are mainly used for metal analysis. It is an ideal ionization source for mass spectrometry, and can ionize over 90% of many elements. Mass spectrophotometer (e.g. ion trap, quadrupole or time-of-flight) covers different mass-to-charge ranges; differ in mass accuracy and achievable resolution.

During ICP-MS process, the ICP source is used to decompose, atomize and ionize a sample of interest. The ions generated in the high temperature argon plasma core are subsequently sorted by mass with the mass spectrophotometer and subjected to further elemental and isotopic analysis. The identities of the ions are determined by their mass-to-charge ratio using a mass analyzer, while the ions intensity is measured at ppt to ppm levels using the ion detector, then the intensity measurements are converted to elemental concentration by comparison with calibration standards. With the high sensitivity and specificity, ICP-MS has been widely used for the detection, characterization, and quantification of nanoparticles [38].

2.2.3. Single particle inductively coupled plasma-mass spectrometry

ICP-MS can be used in single particle mode to characterize individual particles, termed single particle inductively coupled plasma-mass spectrometry (SP-ICP-MS). During SP-ICP-MS process, the sample is first suspended in a nebulized liquid and subsequently carried to argon plasma, where the sample is sequentially desolvated and atomized and ionized, creating a plume of ions. The ions pass through the mass spectrometer where they are separated by mass-to-charge ratio and detected using a time resolved analysis acquisition. The sample solution needs to be diluted sufficiently to ensure low concentrations (ppt to ppb) that no more than one particle will enter the plasma at a time. By using sufficiently short integration (dwell) time which is a duration for the instrument to take a reading, thousands of fast and individual readings are generated to capture nanoparticle event as a discrete signal pulse, each pulse is assumed to correlate to one nanoparticle event [39–41]. Based on ionic calibration standard, the particle mass can be determined by the intensity of the ICP-MS response. If the density of the elemental constituents of the particle is known, the theoretical size of the particle can be determined. If the transport efficiency from the nebulizer to the plasma is known, then the particle number concentration can be further calculated [38, 42].

SP-ICP-MS has been widely applied to measure particle size, size distribution, number concentration and elemental composition of nanoparticles in biological samples, demonstrating it as a powerful tool in quantifying NPs. To deal with biological tissues, a strong acid extraction procedure is required to release the NPs from the matrix. This introduces the possible dissolution of metal NPs which challenges the accuracy of the final analytical data. To solve this problem, Dan et al. studied recoveries of gold NPs when using such a special macerating enzyme that appeared to release the NPs from plant tissue without changing the size distribution of the NPs [43]. With the aid of enzymatic digestion, we have applied SP-ICP-MS analysis to characterize Ag NPs internalized by Arabidopsis, thus having established a new technique and opened up new research domain in our lab [44]. Overview of these analytical techniques including advantages and limitations with examples of application in plant-NP interaction studies is provided in **Table 1**.

3. Detection and characterization of nanoparticles in plants

Although a range of techniques are available to detect and characterize uptake, translocation and biotransformation of NPs in plant tissue, no single technique can provide all information regarding plant-NP interaction. Sufficient information is often obtained by the combination of these analytical techniques, which could provide complementary information mutually. Here in this section literature examples from recent studies are used to demonstrate the application of different techniques in the study of plant-NP interaction.

3.1. Before NPs application

Careful characterization of NPs is critical for accurately assessing the impacts of nanoparticles on plants and understanding their behavior. When initiate an experiment, NPs will either diffuse or aggregate within certain biological media due to different characteristics of the media (i.e. pH, ionic strength, concentration and redox conditions) [45, 46], the aggregation state of NPs will result in quite distinct properties from original NPs. Therefore, the characterization of original NPs is often the first step before NPs application [45–47]. Zhang et al. used TEM images to observe the shape and size of $nCeO_2$ before applied to romaine lettuce. XRD spectrum confirmed the cubic fluorite structure of $nCeO_2$; then ICP-MS was used to confirm the purity of $nCeO_2$. Measuring zeta potential and hydrodynamic size of $nCeO_2$ by DLS analysis indicated a significant aggregation of particles after mixing $nCeO_2$ with nutrient solution. After $nCeO_2$ application, μ-XRF analysis showed that Ce mostly distributed outside the roots. TEM images confirmed that large amount of $nCeO_2$ aggregates distributed on the root surface [48]. Yang et al. used TEM to measure averaged size of CeO_2-NPs suspended in deionized water, and XRD was employed to detect average primary particle size of CeO_2-NPs in dry powder samples, as well as to confirm the crystal structures of CeO_2-NPs [49]. Vinković et al. used DLS, TEM and ICP-MS to characterize AgNPs in ultrapure water (UPW) and sterilized tap water used for the plant watering (TWW) [47]. By measuring hydrodynamic diameter, zeta potential and polydispersity index (PdI) of citrate-coated AgNPs, DLS results showed that the volume size distribution in UPW was bimodal with 90% of smaller particles (12.9 ± 9.1 nm) and only 9% of bigger particles (87.6 ± 41.7 nm). The zeta potential value equal to −16.9 ± 0.6 mV indicated electrostatic stabilization of AgNPs in UPW. While after suspension in TWW, AgNPs aggregation occurred due to higher ionic strength of TWW. Further TEM analysis confirmed the presence of flocculated and aggregated AgNPs in TWW. ICP-MS was used to estimate the stability of AgNPs upon dissolution, results showed that total Ag was lower than 0.5% in TWW, which implies that Ag^+ release was not occurred in TWW [47].

3.2. Uptake and translocation of NPs in plants

In order to understand the uptake mechanism of NPs and their translocation pathway, imaging techniques are often employed to visualize the distribution and morphology of NPs upon exposure, EDS can provide information on their chemical composition, while ICP-based techniques are used to measure particle number concentration, size distribution and mass concentration. Zhao et al. used SEM imaging to find that the root tip of *Eichhornia crassipes* after CuO NP exposure was thinner than unexposed root. EDS analysis of the aggregates attached on epidermis showed the presence of 37.6% (w/w) of Cu, confirmed that CuO NPs presented on the surface of root tips. Further through TEM imaging, dark aggregates with high electron density were detected in the intercellular spaces of cortical tissues in roots, and EDS analysis confirmed the presence of Cu on these aggregates, indicating that CuO NPs were taken up by roots and located in intercellular spaces [50]. Yang et al. used ICP-MS to find that CeO_2-NPs were taken up from root and subsequently translocated to shoot tissues in *Arabidopsis thaliana*, Ce accumulation was much higher in CeO_2-NP treatments than those in CeO_2-bulk and ionic Ce treatments, indicated that the toxicity resulted from the CeO_2-NPs per se rather than from the dissolved Ce ions. TEM images showed the presence of a large number of needle-like particle aggregations in the intercellular regions and the cytoplasm of leaf cells [49]. Stegemeier et al. used synchrotron-based μ-XRF to visualize silver distribution in duckweed roots exposed to Ag^0 NPs or Ag_2S NPs, or to $AgNO_3$. The silver Kα XRF maps showed clear differences in the distribution of Ag for each type of Ag used. The silver was distributed throughout the root tip and showed highest concentrations near the apical meristem after exposure to $AgNO_3$. A similar distribution of Ag in root tip was shown after exposure to Ag^0-NPs. While after exposure to Ag_2S-NPs, a hotspot of silver located at the end of the root cap, suggested that silver was not readily internalized in this case [51]. Pradas del Real et al. firstly used μ-CT to create 3D reconstructed image of wheat root after Ag NPs exposure for in situ 3D visualization, then μ-XRF was used to provide 2D elemental distribution [52]. Combination of μ-CT and μ-XRF showed the presence of localized Ag accumulation regions with a size of 1–4 μm adhering on the epidermis. Nano-CT technique capable of higher spatial resolution revealed that these

AgNPs accumulated preferentially in discontinuities between root epidermal cells. In addition, many AgNPs were fixed on root hairs. With the methods to study Ag_2S-NPs treatment, μ-XRF showed that Ag is mainly colocalized with S, μ-CT and nano-CT showed that these Ag accumulation regions with a size from 3 to 8 μm presented mostly on the root surface. Through ICP-MS analysis, a higher Ag content in root and shoot was observed after exposure to AgNPs compared to $AgNO_3$ exposure, suggesting a nano-specific accumulation mechanism [52]. Peng et al. used ICP-OES to measure Cu content in root of rice after adding CuO NPs to the soil. The results showed that Cu content in roots was significantly increased, with a much higher content than aboveground parts. μ-XRF analysis indicated that Cu accumulated in the aleuronic layer of rice, but not the polished rice [53]. In another study, In order to study whether CeO_2 NPs can move from the roots to shoots in cucumber after the root was exposed to CeO_2 NPs, TEM and EDS analyses were performed and the presence of Ce particles in the xylem sap was confirmed, suggested that Ce-containing species could be transported throughout the whole plant by vascular system. ICP-MS analysis also confirmed the uptake of CeO_2 NPs from root to shoot [54]. Li et al. used the macerozyme R-10 tissue extraction method followed by SP-ICP-MS to study the uptake and size distribution of AgNPs in soybean and rice. Both SP-ICP-MS and TEM measurements indicated that the size of Ag-containing NPs were 2–3 times larger than the originally dosed AgNPs after exposure to AgNPs, indicating the AgNPs biotransformation processes were involved [55].

3.3. Biotransformation of NPs in plants

Biotransformation is defined as biochemical modification by living organisms [16, 56]. Biotransformation of NPs by plant may modify the toxicity, behavior, and fate of NPs in the plant tissue. Biotransformation process may involve redox, dissolution, sulfidation, aggregation, and adsorption of macromolecules and ion [10, 57]. XAS and STXM are the most frequently used techniques to characterize the speciation of NPs during cellular internalization and intracellular biotransformation. Zhao et al. used EDS technique to find that S present on aggregates in the intercellular spaces of cortical tissues in *Eichhornia crassipes* (water hyacinth) roots after CuO NP exposure, indicating that CuO NPs (or other Cu species) interacted with S-containing compounds such as cysteine. XANES was employed to identify Cu species in roots and leaves after CuO NPs internalization. XANES analysis revealed that CuO NPs in roots mainly kept the original pattern (65.7% of CuO). Other Cu species included Cu-Ac (14.2%), $Cu_2(OH)PO_4$ (8.7%) and 7.6% of Cu_2S. XRD spectrum of original CuO NPs showed that all peaks belonged to CuO, and no peak on Cu_2S was detected, indicating that the observed Cu_2S in roots were formed after incubation with CuO NPs [50]. Zhang et al. used Bulk-XANES technique to study transformation of $nCeO_2$. XANES spectra of root and shoot showed similar feature as the initial $nCeO_2$; Results showed that Ce in lettuce mostly presented as CeO_2, with a small fraction of $CePO_4$ in roots (4.3%) and Ce carboxylates (3.5%) in leaves, suggested that $nCeO_2$ can release small amount of Ce^{3+} with the assistance of organic acids and reducing substances in root exudates [48, 58]. Stegemeier et al. used EXAFS spectra to determine Ag speciation in duckweed (*Landoltia punctata*) roots after exposure to Ag^0 NPs or Ag_2S NPs, or $AgNO_3$, revealed that more photo-reducible Ag species were generated after exposure to ionic Ag [51]. In contrast, a higher prevalence of sulfur associated Ag species (as a mixture of Ag_2S (64%) and Ag-thiol (53%) were produced after exposure to Ag^0 NPs or Ag_2S NPs treatment. Bulk EXAFS analysis of Ag_2S-NP treatment indicated that plant is unable to dissolve or transform a significant amount of the Ag_2S-NPs after 24 h exposure [51]. In another study, μ-XANES spectroscopy was employed to determine speciation of Ag at root after Ag NPs exposure.

μ-XANES revealed that Ag was mostly present as metallic Ag in the epidermis, but inside the roots Ag was homogeneously distributed in the cell walls of the cortex as a mixture of Ag-thiol species and other ionic Ag species, suggested the biotransformation of Ag occurred. Moreover, no Ag(0) was observed inside roots, implied that Ag-NPs were completely dissolved and complexed by organic ligands [52]. Peng et al. used Bulk-XANES to analyze translocation and transformation of CuO NPs in rice, the results revealed that Cu element mainly existed in the form of copper citrate, only a small portion of Cu kept original CuO form in roots, stems, and leaves of rice after CuO NP treatment. During CuO NPs internalization in rice, one-third of Cu(II) was transformed to Cu(I) which was mainly associated with cysteine. CuO, copper citrate, and copper (I) acetate all accounted for nearly 30% of the total Cu in the chaff [53]. Ma et al. combined μ-XRF and μ-XANES to detect CeO_2 NPs or its transformation species in the xylem sap, shoots and roots of cucumber after exposure to CeO_2 NPs, revealed that about 15% of Ce was reduced from Ce(IV) to Ce(III) in the roots after treatment, and Ce was transported as a mixture of Ce(IV) and Ce(III) from roots to shoots through xylem, while was transported almost only in the form of CeO_2 from shoots back to roots through phloem [54]. Peng et al. used bulk-XANES to analyze Cu speciation in the tissues of rice plants after exposure to CuP NPs, indicated that Cu was combined with cysteine, citrate, and phosphate ligands, and some of the Cu (II) was transformed to Cu (I) during CuO NP uptake, and confirmed that CuO NPs were transported from the roots to the leaves. In order to further study Cu biotransformation in cellular level, they firstly used μ-XRF to map Cu element distribution in the root; the results revealed that Cu was mainly localized in the root epidermis and exodermis. Then μ-XANES was employed to determine speciation of Cu element at selected spots in μ-XRF map. In addition, combination of STXM with Cu L3-edge XANES spectroscopy was used to map the in situ elemental composition of Cu in the root cells, the results confirmed that speciation of Cu in the root cells and the intercellular space existed in the form of Cu-citrate and CuO NPs, respectively [10]. Zhang et al. used TEM to detect the uptake and localization of nano-Yb_2O_3 in cucumber roots after exposure, found that a lot of high electron-dense dark deposits looked like fine needle-shaped nanoclusters in the intercellular spaces and middle lamellas in the cross sections of cucumber roots, later EDS analysis confirmed the presence of Yb in these dark deposits. In order to identify the chemical species of Yb in these dark deposits, the chemical distribution was mapped by STXM, and NEXAFS spectra were extracted, results indicated that this compound was inferred to be $YbPO_4$, suggesting that Yb_2O_3 particles and $YbCl_3$ were all transformed to $YbPO_4$ in the intercellular regions of the roots, and indicating that biotransformation and internalization of Yb_2O_3 nanoparticle took place in plant cell, which conferred phytotoxicity to plant [59]. In another study, Zhang et al. used the same methodology to investigate the biotransformation of CeO_2 NPs in cucumber. TEM images showed the presence of needle-like clusters on the epidermis and in the intercellular spaces of cucumber roots after CeO_2 NPs exposure. STXM imaging indicated that the chemical composition of needle-like clusters is $CePO_4$. Further XANES analysis showed that Ce presented in the roots as CeO_2 and $CePO_4$ while in the shoots as CeO_2 and cerium carboxylates, confirming biotransformation of CeO_2 NPs in plant cells [16]. In order to determine the toxicity and fate of nanoparticles upon exposure to plants, Wang et al. combined a variety of techniques to investigate the cellular internalization and intracellular biotransformation of silver nanoparticles in *Chlamydomonas reinhardtii*. NanoSIMS was firstly applied to analyze the distributions of Ag in algal cells, silver was observed to accumulate predominantly on the cell walls and in the cytoplasm of the algae after exposure to AgNPs. Then HAADF-STEM was performed to examine the accurate localization and morphology of Ag, HAADF image showed that a set of bright spots

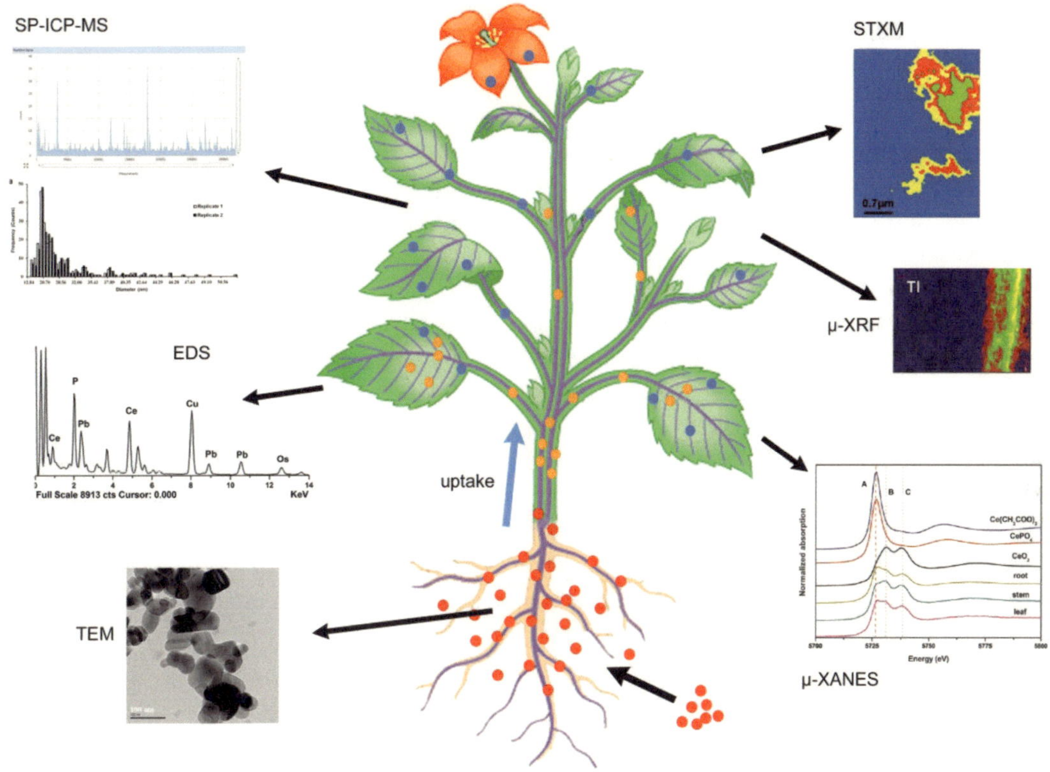

Figure 1. Schematic diagram represent a regular work flow of NPs characterization in plant. A selection of analytical techniques is shown. Red dots indicate NPs at the moment of application. Yellow and blue dots indicate different elemental species of NPs after biotransformation in plants. The images of SP-ICP-MS, EDS, TEM, STXM, µ-XRF and µ-XANES are adapted from [16, 18, 44, 60].

located mainly in the periplasmic space and cytoplasm, TEM was further used to observe morphology of these bright spots. EDS analysis showed that these bright spots were Ag-containing substances; moreover, Ag and S always occurred concomitantly. EDS-mapping confirmed that Ag was almost exclusively co-localized with S in the cytoplasm of algae but not in the periplasmic space. Later, Synchrotron based Ag K-edge XAS was performed to further identify Ag speciation after exposure. It was found that Ag glutathione complexes and Ag_2S represented the main speciation, suggested that Silver was also found to coexist with sulfur inside the cytoplasm in the form of Ag-GSH and Ag_2S [36]. A regular work flow of NPs characterization during plant-NP interaction with the application of the most-frequently used techniques is shown in **Figure 1**.

4. Conclusion and future perspectives

Although the combination of these techniques described in this review is capable of taking over most of the task on the characterization of NPs during plant-NP interaction, considerable limitations of these techniques still remain to overcome. Many techniques are destructive, such as TEM, SEM and nanoSIMS, which means the same sample cannot be analyzed twice or by another method for validation. Analytical artifacts are sometimes inevitable during some sample preparation procedures. Because biological samples is usually hetero-dispersed and multicomponent, with diverse elemental compositions and sometimes contain multiple types of NPs, the analysis of NPs in these samples is thus quite complicated and a variety of methodology is utilized to provide complementary information, while the results measured by these different methods are not always comparable, which may partially due to different sample preparation procedures in different techniques. Further, instrument operation procedures and statistical analyses are likely to contribute to the complexity and uncertainty. Another challenge arises when analyze samples with low concentrations of the analyte. In non-hyperaccumulating plant species, visualizing the spatial distribution of NPs and detecting reliable in situ information about the chemical speciation of trace elements will be very difficult. In this case, analytical techniques with high sensitivity are desired to measure low concentrations of NPs. An ideal analytical technique should be able to simultaneously determine all parameters regarding plant-NP interaction, such as article size, morphology, structure, size distribution, mass concentration, translocation, elemental speciation and etc. It should be sensitive and accurate enough for in situ detection and characterization of trace element in complex biological samples in a non-destructive way. Although none of the existing techniques are able to solely provide all the information desired, we believe that a promising evolution of analytical methodology is taking place and will be capable of fulfilling requirements as much as desired to provide sufficient information about plant-NP interaction.

Acknowledgements

Authors thank the NIE AcRF grant (RI 8/16 CZ).

Conflict of interest

Authors declare no conflict of interest.

Abbreviations

DLS	dynamic light scattering
EDS	energy-dispersive X-ray spectroscopy
EXAFS	extended X-ray absorption fine structure
ICP-MS	inductively coupled plasma-mass spectrometry
ICP-OES	inductively coupled plasma-optical emission spectrometry
NanoSIMS	nano secondary ion mass spectrometry
NPs	nanoparticles
SEM	scanning electron microscope
SP-ICP-MS	single particle-inductively coupled plasma-mass spectrometry

STEM	scanning transmission electron microscope
STXM	scanning transmission X-ray microscopy
TEM	transmission electron microscope
XANES	X-ray absorption near edge structure
XAS	X-ray absorption spectrometry
XRD	X-ray diffraction
XRF	X-ray fluorescence spectrometry
μCT	X-ray computed microtomography

Author details

An Yan and Zhong Chen*

*Address all correspondence to: zhong.chen@nie.edu.sg

Natural Sciences and Science Education Academic Group, National Institute of Education, Nanyang Technological University, Singapore

References

[1] Sekhon BS. Nanotechnology in agri-food production: An overview. Nanotechnology, Science and Applications. 2014;**7**:31-53

[2] Nel A, Xia T, Mädler L, Li N. Toxic potential of materials at the nanolevel. Science. 2006;**311**:622-627

[3] Zhu H, Han J, Xiao JQ, Jin Y. Uptake, translocation, and accumulation of manufactured iron oxide nanoparticles by pumpkin plants. Journal of Environmental Monitoring. 2008;**10**:713-717

[4] Handy RD, Owen R, Valsami-Jones E. The ecotoxicology of nanoparticles and nanomaterials: Current status, knowledge gaps, challenges, and future needs. Ecotoxicology. 2008;**17**:315-325

[5] Ma X, Geiser-Lee J, Deng Y, Kolmakov A. Interactions between engineered nanoparticles (ENPs) and plants: Phytotoxicity, uptake and accumulation. Science of the Total Environment. 2010;**408**:3053-3061

[6] Wang P, Lombi E, Zhao F-J, Kopittke PM. Nanotechnology: A new opportunity in plant sciences. Trends in Plant Science. 2016;**21**:699-712

[7] Dietz K-J, Herth S. Plant nanotoxicology. Trends in Plant Science. 2011;**16**:582-589

[8] Miralles P, Church TL, Harris AT. Toxicity, uptake, and translocation of engineered nanomaterials in vascular plants. Environmental Science & Technology. 2012;**46**:9224-9239

[9] Judy JD, Bertsch PM. Chapter one—Bioavailability, toxicity, and fate of manufactured nanomaterials in terrestrial ecosystems. In: Sparks DL, editor. Advances in Agronomy. United States: Academic Press; 2014. pp. 1-64

[10] Peng C, Duan D, Xu C, Chen Y, Sun L, et al. Translocation and biotransformation of CuO nanoparticles in rice (*Oryza sativa* L.) plants. Environmental Pollution. 2015;**197**:99-107

[11] Stampoulis D, Sinha SK, White JC. Assay-dependent phytotoxicity of nanoparticles to plants. Environmental Science & Technology. 2009;**43**:9473-9479

[12] Auffan M, Rose J, Bottero J-Y, Lowry GV, Jolivet J-P, et al. Towards a definition of inorganic nanoparticles from an environmental, health and safety perspective. Nature Nanotechnology. 2009;**4**:634-641

[13] Gardea-Torresdey JL, Rico CM, White JC. Trophic transfer, transformation, and impact of engineered nanomaterials in terrestrial environments. Environmental Science & Technology. 2014;**48**:2526-2540

[14] Geisler-Lee J, Brooks M, Gerfen J, Wang Q, Fotis C, et al. Reproductive toxicity and life history study of silver nanoparticle effect, uptake and transport in *Arabidopsis thaliana*. Nanomaterials. 2014;**4**:301

[15] Montes A, Bisson MA, Gardella JA, Aga DS. Uptake and transformations of engineered nanomaterials: Critical responses observed in terrestrial plants and the model plant *Arabidopsis thaliana*. Science of the Total Environment. 2017;**607-608**:1497-1516

[16] Zhang P, Ma Y, Zhang Z, He X, Zhang J, et al. Biotransformation of ceria nanoparticles in cucumber plants. ACS Nano. 2012;**6**:9943-9950

[17] Chau C-F, Wu S-H, Yen G-C. The development of regulations for food nanotechnology. Trends in Food Science & Technology. 2007;**18**:269-280

[18] Tiede K, Boxall ABA, Tear SP, Lewis J, David H, et al. Detection and characterization of engineered nanoparticles in food and the environment. Food Additives & Contaminants: Part A. 2008;**25**:795-821

[19] Stokes D. Principles and Practice of Variable Pressure: Environmental Scanning Electron Microscopy (VP-ESEM). United States: John Wiley & Sons; 2008

[20] Akbari B, Pirhadi Tavandashti M, Zandrahimi M. Particle size characterization of nanoparticles—A practical approach. Iranian Journal of Materials Science & Engineering. 2011;**8**:48-56

[21] Laborda F, Bolea E, Cepriá G, Gómez MT, Jiménez MS, et al. Detection, characterization and quantification of inorganic engineered nanomaterials: A review of techniques and methodological approaches for the analysis of complex samples. Analytica Chimica Acta. 2016;**904**:10-32

[22] Hendrickson OD, Safenkova IV, Zherdev AV, Dzantiev BB, Popov VO. Methods of detection and identification of manufactured nanoparticles. Biophysics. 2011;**56**:961-986

[23] Liu J. Scanning transmission electron microscopy and its application to the study of nanoparticles and nanoparticle systems. Journal of Electron Microscopy. 2005;**54**:251-278

[24] Kim B, Park C-S, Murayama M, Hochella MF. Discovery and characterization of silver sulfide nanoparticles in final sewage sludge products. Environmental Science & Technology. 2010;**44**:7509-7514

[25] Brar SK, Verma M. Measurement of nanoparticles by light-scattering techniques. TrAC Trends in Analytical Chemistry. 2011;**30**:4-17

[26] Goldstein JI, Newbury DE, Michael JR, Ritchie NW, Scott JHJ, et al. Scanning Electron Microscopy and X-Ray Microanalysis. Germany: Springer; 2017

[27] Lombi E, Susini J. Synchrotron-based techniques for plant and soil science: Opportunities, challenges and future perspectives. Plant and Soil. 2009;**320**:1-35

[28] Gräfe M, Donner E, Collins RN, Lombi E. Speciation of metal(loid)s in environmental samples by X-ray absorption spectroscopy: A critical review. Analytica Chimica Acta. 2014;**822**:1-22

[29] Zhao F-J, Moore KL, Lombi E, Zhu Y-G. Imaging element distribution and speciation in plant cells. Trends in Plant Science. 2014;**19**:183-192

[30] Ortega R, Carmona A, Llorens I, Solari PL. X-ray absorption spectroscopy of biological samples: A tutorial. Journal of Analytical Atomic Spectrometry. 2012;**27**:2054-2065

[31] Manceau A, Boisset M-C, Sarret G, Hazemann J-L, Mench M, et al. Direct determination of lead speciation in contaminated soils by EXAFS spectroscopy. Environmental Science & Technology. 1996;**30**:1540-1552

[32] Landis EN, Keane DT. X-ray microtomography. Materials Characterization. 2010;**61**:1305-1316

[33] de Groot FMF, de Smit E, van Schooneveld MM, Aramburo LR, Weckhuysen BM. In-situ scanning transmission X-ray microscopy of catalytic solids and related nanomaterials. Chemphyschem. 2010;**11**:951-962

[34] Bluhm H, Andersson K, Araki T, Benzerara K, Brown GE, et al. Soft X-ray microscopy and spectroscopy at the molecular environmental science beamline at the advanced light source. Journal of Electron Spectroscopy and Related Phenomena. 2006;**150**:86-104

[35] Herrmann AM, Ritz K, Nunan N, Clode PL, Pett-Ridge J, et al. Nano-scale secondary ion mass spectrometry—A new analytical tool in biogeochemistry and soil ecology: A review article. Soil Biology and Biochemistry. 2007;**39**:1835-1850

[36] Wang S, Lv J, Ma J, Zhang S. Cellular internalization and intracellular biotransformation of silver nanoparticles in *Chlamydomonas reinhardtii*. Nanotoxicology. 2016;**10**:1129-1135

[37] Moore KL, Lombi E, Zhao F-J, Grovenor CRM. Elemental imaging at the nanoscale: NanoSIMS and complementary techniques for element localisation in plants. Analytical and Bioanalytical Chemistry. 2012;**402**:3263-3273

[38] Wilbur S, Yamanaka M, Sannac S. Characterization of Nanoparticles in Aqueous Samples by ICP-MS. United States: Agilent Publication; 2015

[39] Mitrano D, Ranville J, Stephan C, Shelton C (2014) Quantitative Evaluation of Nanoparticle Dissolution Kinetics Using Single Particle ICP-MS: A Case Study with Silver Nanoparticles. PerkinElmer application note

[40] Mitrano DM, Lesher EK, Bednar A, Monserud J, Higgins CP, et al. Detecting nanoparticulate silver using single-particle inductively coupled plasma-mass spectrometry. Environmental Toxicology and Chemistry. 2012;**31**:115-121

[41] Pace HE, Rogers NJ, Jarolimek C, Coleman VA, Gray EP, et al. Single particle inductively coupled plasma-mass spectrometry: A performance evaluation and method comparison in the determination of nanoparticle size. Environmental Science & Technology. 2012;**46**:12272-12280

[42] Pace HE, Rogers NJ, Jarolimek C, Coleman VA, Higgins CP, et al. Determining transport efficiency for the purpose of counting and sizing nanoparticles via single particle inductively coupled plasma mass spectrometry. Analytical Chemistry. 2011;**83**:9361-9369

[43] Dan Y, Zhang W, Xue R, Ma X, Stephan C, et al. Characterization of gold nanoparticle uptake by tomato plants using enzymatic extraction followed by single-particle inductively coupled plasma–mass spectrometry analysis. Environmental Science & Technology. 2015;**49**:3007-3014

[44] Bao D, Oh ZG, Chen Z. Characterization of silver nanoparticles internalized by *Arabidopsis* plants using single particle ICP-MS analysis. Frontiers in Plant Science. 2016;**7**:32

[45] Vinkovic Vrcek I, Pavicic I, Crnkovic T, Jurasin D, Babic M, et al. Does surface coating of metallic nanoparticles modulate their interference with in vitro assays? RSC Advances. 2015;**5**:70787-70807

[46] Auffan M, Rose J, Wiesner MR, Bottero J-Y. Chemical stability of metallic nanoparticles: A parameter controlling their potential cellular toxicity in vitro. Environmental Pollution. 2009;**157**:1127-1133

[47] Vinković T, Novák O, Strnad M, Goessler W, Jurašin DD, et al. Cytokinin response in pepper plants (*Capsicum annuum* L.) exposed to silver nanoparticles. Environmental Research. 2017;**156**:10-18

[48] Zhang P, Ma Y, Liu S, Wang G, Zhang J, et al. Phytotoxicity, uptake and transformation of nano-CeO_2 in sand cultured romaine lettuce. Environmental Pollution. 2017;**220**:1400-1408

[49] Yang X, Pan H, Wang P, Zhao F-J. Particle-specific toxicity and bioavailability of cerium oxide (CeO_2) nanoparticles to *Arabidopsis thaliana*. Journal of Hazardous Materials. 2017;**322**:292-300

[50] Zhao J, Ren W, Dai Y, Liu L, Wang Z, et al. Uptake, distribution, and transformation of CuO NPs in a floating plant *Eichhornia crassipes* and related stomatal responses. Environmental Science & Technology. 2017;**51**:7686-7695

[51] Stegemeier JP, Colman BP, Schwab F, Wiesner MR, Lowry GV. Uptake and distribution of silver in the aquatic plant *Landoltia punctata* (duckweed) exposed to silver and silver sulfide nanoparticles. Environmental Science & Technology. 2017;**51**:4936-4943

[52] Pradas del Real AE, Vidal V, Carrière M, Castillo-Michel H, Levard C, et al. Silver nanoparticles and wheat roots: A complex interplay. Environmental Science & Technology. 2017;**51**:5774-5782

[53] Peng C, Xu C, Liu Q, Sun L, Luo Y, et al. Fate and transformation of CuO nanoparticles in the soil–rice system during the life cycle of rice plants. Environmental Science & Technology. 2017;**51**:4907-4917

[54] Ma Y, He X, Zhang P, Zhang Z, Ding Y, et al. Xylem and phloem based transport of CeO_2 nanoparticles in hydroponic cucumber plants. Environmental Science & Technology. 2017;**51**:5215-5221

[55] Li C-C, Dang F, Li M, Zhu M, Zhong H, et al. Effects of exposure pathways on the accumulation and phytotoxicity of silver nanoparticles in soybean and rice. Nanotoxicology. 2017;**11**:699-709

[56] Abramowicz DA. Aerobic and anaerobic biodegradation of PCBs: A review. Critical Reviews in Biotechnology. 1990;**10**:241-251

[57] Lowry GV, Gregory KB, Apte SC, Lead JR. Transformations of nanomaterials in the environment. Environmental Science & Technology. 2012;**46**:6893-6899

[58] Zhang P, Ma Y, Zhang Z, He X, Li Y, et al. Species-specific toxicity of ceria nanoparticles to *Lactuca* plants. Nanotoxicology. 2015;**9**:1-8

[59] Zhang P, Ma Y, Zhang Z, He X, Guo Z, et al. Comparative toxicity of nanoparticulate/bulk Yb_2O_3 and $YbCl_3$ to cucumber (*Cucumis sativus*). Environmental Science & Technology. 2012;**46**:1834-1841

[60] Scheckel KG, Lombi E, Rock SA, McLaughlin MJ. In vivo synchrotron study of thallium speciation and compartmentation in *Iberis intermedia*. Environmental Science & Technology. 2004;**38**:5095-5100

Chapter 7

Nanotechnology in Agriculture: New Opportunities and Perspectives

Luca Marchiol

Additional information is available at the end of the chapter

http://dx.doi.org/10.5772/intechopen.74425

Abstract

The prediction that in 2050 our planet will be populated by over 9 billion people is quite reliable. This will pose serious problems with food, water and energy supply, particularly in less-developed countries. Considering that the human pressure over natural resources has already reached critical levels, international agencies such as the World Bank and UN Food and Agriculture Organization (FAO) are soliciting scientific research in order to identify innovative solutions to support the primary sector. Nanotechnology is a rapidly evolving field with the potential to take forward the agriculture and food industry with new tools which promise to increase food production in a sustainable manner and to protect crops from pests. Such expectations are coupled with some uncertainties about the fate of nanomaterials in the agro-environment. However, the field application of engineered nanomaterials (ENMs) has not been properly investigated yet, and many aspects have only been considered theoretically or with models, which make it difficult to properly assess the usefulness of ENMs for plant fertilization and protection.

Keywords: agriculture, engineered nanomaterials, plant nanobionics, nanofertilizers, agricultural residues

1. Introduction

The current world population of 7.6 billion is expected to reach 8.6 billion in 2030, 9.8 billion in 2050 and 11.2 billion in 2100 [1]. This implies that new systems for food, water and energy will be needed to ensure food security. On the other hand, producing more food requires natural resources, land consumption, water supply and energy [2]. Thus, in the very near future, scientific research will be requested to provide new paradigms and practices to solve highly complex and diverse problems. Some examples are the following: (i) How will we feed our children? (ii) How can we simultaneously deliver increased crop yields and reduce the environmental impact of agriculture? (iii) How do plants contribute to the ecosystem services (e.g., photosynthesis, nitrogen fixation, and organic matter cycle) upon which humanity depends? Will world agricultural systems be able to cope with global climate change? [3].

Agriculture uses inefficiently the conventional inputs (land, water, energy, fertilizers and pesticides), and a large fraction of plant protection products applied per year are lost or are unavailable to the target [4, 5]. In addition, agriculture (cultivation of crops, livestock and deforestation) is a major contributor to greenhouse gas emissions producing about 24% of the total annual worldwide amount [6]. Waste production is another relevant issue of the primary sector. European countries produce approximately 90 million tons of agricultural wastes per year [7]. Nanotechnology has been recognized by the European Commission as one of its six "Key Enabling Technologies" that contributes to sustainable competitiveness and growth in several fields of industrial applications underpinning the shift to a greener economy [8]. Before beginning to deepen the analysis of the potential benefits of applying nanoscience to agriculture, we have to answer the following question: Why and how are nanotechnology and engineered nanomaterials (ENMs) expected to respond to the abovementioned issues?

Specific answers are provided by recent scientific literature which reports promising opportunities for nanoscience and nanotechnology to improve sustainability of agri-food systems [9]. From a quantitative perspective, by examining the growth of scientific literature on nanotechnology, it appears clearly that the interest on research in this field grew significantly between the end of twentieth century and the beginning of twenty-first century [10].

Compared to other fields of nanotechnology application, like medicine, materials and energy, agriculture is still a marginal sector. However, publications dedicated to agricultural applications tend to increase similar to those observed in other sectors. This is demonstrated by the increasing number of peer-reviewed scientific literature per year retrieved in Elsevier Scopus database. The query "Nanotechnology" and "Agriculture" was launched last October 10, 2017. The results were limited to the period 2000–2017 and filtered for scientific papers, reviews and conference papers. A number of 508 scientific products have been indexed in Scopus database: 264 (52%) papers, 143 (285%) reviews and 91 (18%) conference papers. As regards the distribution of publications among the most productive countries, United States and India share the research leadership in this field, having published together about 63% of papers (27 and 25%, respectively). China possesses the third rank (10%), whereas EU countries contribute with about 20% of publications.

The unique physicochemical properties of nanomaterials, that is, catalytic reactivity, high surface area, size and shape, have the potential to open new paradigms and to introduce new strategies in agriculture. Such new paradigms request also new terms. In a recently published book, the term "Agri-nanotechniques" was used. Since no specific definition for this word was provided, it has been used to indicate nanosystems utilized for the delivery of nutrient elements in crops [11]. In more general terms, the application of nanotechnology in the plant production systems or—more broadly—in plant science was defined with the term "Phytonanotechnology" [12]. However, since nanotechnology application in agriculture is in its infancy, it is very likely that new words will be invented in some time to indicate more specific technical developments.

The state-of-the-art R&D of nanotechnology for the agricultural sector and their potential market in EU were firstly analyzed in 2013 during the workshop on "Nanotechnology for the agricultural sector: from research to the field" organized by the JRC-IPTS [13]. More recently, the European Food Safety Authority (EFSA) provides an inventory of current and potential future applications of nanotechnology in the agri-food sector and to review the regulation of nanomaterials in the EU as well as in non-EU countries [14].

So far, we have discussed in general about food security and the agricultural sector. In the next paragraphs, we will narrow down the analysis of nanotechnology applications specifically dedicated to field crop and plant production. Other sectors such as plant protection, animal husbandry and food technology will be not considered.

Specific agronomic applications of nanotechnology include (i) enabled delivery systems of release of agrochemicals allowing a controlled release of fertilizers, pesticides and herbicides, (ii) field-sensing systems to monitor the environmental stresses and crop conditions and (iii) improvement of plant traits against environmental stress and diseases [15, 16].

2. New opportunities

There are at least two fundamental aspects in the management of primary production on which research can produce significant advances to meet future needs: (i) increased production rate and crop yield, (ii) increased efficiency of resource utilization and (iii) reduction of waste production.

2.1. Increase production rate and crop yield

Crop yield increases have been achieved by utilizing plant breeding, fertilizers and plant-protection-products [17]. Since Green Revolution, which occurred during the decade 1960–1970, agricultural productivity growth has been in decline and at present we need a second revolution in agricultural technology [18]. However, rather than an increase in the doses of traditional agronomic factors, it is realistic that significant improvements in crop yield will come from improving the efficiency of the photosynthetic process.

Food security is based on plant photosynthesis. About 85% of plant species are C3 plants which are the most common and efficient in photosynthesis in cool wet climates. They include the cereal grains: wheat, rice, barley, oats, cotton, sugar beets, tobacco and soybean. In addition, most trees and most lawn grasses such as rye and fescue are C3 plants.

Photosynthetic organisms are able to convert radiant energy from solar light into chemical energy which is stored in sugars. The process coupled biophysical processes—absorption of photosynthetically active radiation (PAR) and electron transport—and biochemical processes—NADPH and ATP. Some targets have been identified to improve the photosynthesis [19].

Among these, the most serious candidate is the photosynthetic enzyme ribulose-1,5-bisphosphate carboxylase/oxygenase—in short, Rubisco. This molecule catalyzes the addition of CO_2 to the five-carbon compound ribulose bisphosphate, in the initial phase of the Calvin-Benson cycle [20]. Rubisco also reacts with oxygen in photorespiration. This is considered a wasteful process; in fact, it was verified that in C3 plants (25°C, current atmospheric [CO_2]), about 30% of fixed C is lost to recover Rubisco. For that reason, Rubisco is considered the physiological "bottleneck" of photosynthesis [21].

Let us take a step back and reconsider the biophysical processes of photosynthesis. More precisely, we take into consideration the energy source that promotes the process, that is, solar radiation. Visible light corresponds to 43% of solar light; it lies between 400 and 700 nm in the solar spectrum and approximately coincides with PAR. When sunlight reaches the leaf surface the photosynthetic pigments chlorophyll-a and chlorophyll-b absorb photons as allowed by their absorption spectrum and provide the energy to the biochemical pathway of photosynthesis [22]. The process is highly inefficient, the solar energy conversion efficiency (ratio of the energy stored to the energy of light absorbed) being 2.4 and 3.7%, respectively, for C3 and C4 healthy crops [23].

2.1.1. *Plant nanobionics and photosynthesis*

For years, important discussions and studies are under way to fill the knowledge gaps in order to overcome the limitation of photosynthesis. Significant efforts are made working on different strategies, including (i) engineering C3 crops to use C4 photosynthesis pathway [24], (ii) improving the efficiency of Rubisco [25], (iii) modifying the chlorophyll antenna size of chloroplast photosystems [26], (iv) improving the recovery rate from photoinhibition [27] and broadening the photosynthetic light waveband [28]. According to Evans, "recent technological developments now provide us with the means to engineer changes to photosynthesis that would not have been possible previously" [28].

There is no doubt whatsoever that nanotechnology is among these new tools. The scientific literature devoted to the relationships between plants and nanomaterials is not very large yet. However, a relatively large body of papers reported the positive effects of nanomaterials on photosynthesis. Early studies considered titanium oxide nanoparticles ($n$$TiO_2$). And that is because the high photocatalytic activity of anatase crystal $n$$TiO_2$ was hypothesized to have a role in the improvement of light absorbance by plant leaves, thus sustaining an increase in photosynthesis. In particular, it was demonstrated that $n$$TiO_2$ protects the chloroplast from aging due to photochemical stress [29–30], activates Rubisco carboxylation promoting an enhancement of the photosynthetic rate [31–33] and positively influences biophysics traits of photosynthesis, such as electron chain transport and Chl-photophosphorylation activity [34]. Finally, in addition to photosynthesis, $n$$TiO_2$ improves leaf water conductance and transpiration rate [35].

More recently, the original idea to merge nanomaterials with living plants to enhance their native functions and to give them non-native functions has been more accurately focused. This approach assumed the name of "plant nanobionics" [36] and potentially allowed to engineer faster-growing plants and become the key factor to design and develop artificial photosynthetic systems, a potential source of clean energy [37, 38]. In addition, it could also lead to other innovations that we cannot imagine at this time.

The first report demonstrating an application of plant nanobionics was provided by a research group from MIT. A suspension of single-walled carbon nanotubes (SWCNTs) was supplied by perfusion to leaves of *Arabidopsis thaliana* and to isolated chloroplasts of *Spinacia oleracea*. In both cases the SWCNTs were observed within the thylakoids and no symptoms of stress were recorded. The treatment increased the electron transport rate compared to control and the shelf life of isolated chloroplasts was extended by about 2 h. The authors proposed that the semi-conductor SWCNTs have a high electrical conductance and are able to capture solar energy in wavelengths that are weakly absorbed by chloroplasts. In particular, an enhancement in the light absorption profile of chloroplasts by increasing the light energy capture in UV and N-IR ranges of the spectrum was supposed [36].

In their experimental conditions the authors observed that SWCNT-chloroplast assemblies promoted over three times a higher photosynthetic activity than control and enhanced electron transport rate. On the one hand, there is no doubt that still extensive research would be needed to see the effects of plant nanobionics in terms of increased production of sugars as well as crop yield. On the other hand, the enhancement of a basic plant function in response to incorporation of nanomaterials was demonstrated as proof of concept [36].

2.2. Increase in efficiency of resource utilization

2.2.1. Principles of plant nutrition and fertilization

Optimal crop nutrition is a fundamental requirement for food security, which means that fertilization has a prominent role in modern agriculture. Crop yield is highly dependent on macronutrients (N, P, K, S, Ca, Mg) and micronutrients (B, Fe, Mn, Cu, Zn, Mo, and Cl) input to agricultural lands [39]. A conservative estimate obtained by examining the results of a number of long-term field studies on crop production suggested that from 30 to 50% of crop yield is attributable to commercial fertilizer nutrient inputs [40].

Nutrient use efficiency (NUE) is a measure of how well plants use the available mineral nutrients. In all agroecosystems NUE of crop plants is lower than 50% due to physical and chemical soil properties, leaching, gaseous losses and fertilizer characteristics [41], this is, for instance, in the case of urea [CO(NH2)2] which is one of the most important N-fertilizers (46% N by weight). Plants are not able to take up this molecule but the byproducts produced in soil after urea decomposition due to hydrolysis, volatilization and urease soil enzyme [42]. If ammonia is not readily assimilated by plant roots, then, large amounts of nitrogen are lost.

Since the fertilizer use between 1950 and 2000 increased about 20-fold and 7-fold for N and P, respectively [43], we have a 2-fold consequence. On one side, the lower efficiency of fertilizer dose implies that to maintain high production the production costs are increasing. From one another we have risks of environmental pollution.

As for micronutrients, though they are present in plants in concentrations generally below 100 ppm, they play fundamental physiological roles in plant metabolism, being activators of specific enzymes. Many micronutrients stimulate or are part of plant defensive systems against diseases or abiotic stress [44]. Moreover, plants are the sources of these essential elements for animals and humans [45]. Soil micronutrient deficiencies or insufficient micronutrient availability in soils limit crop productivity and nutritional value of food.

The most common method of micronutrient application for crops is soil application. Under unfavorable conditions (neutral to alkaline soil pH) microelements frequently precipitate and become less bioavailable [46]. It has been reported that the fertilizer-micronutrient use efficiency by crops is lower than 5% [47].

To overcome the soil limiting factors, a second strategy widely used to provide micronutrients to crops is via leaf treatments. However, plants primarily absorb nutrients through their roots. The amount of micronutrients that can be absorbed by leaves is limited, and they are not transported to the roots via the phloem (basipetal flux) [48].

2.2.2. Smart fertilizers for crop nutrition

Best management practices for fertilization are those that support the achievement of the main objectives of sustainable agriculture: productivity, profitability and environmental health. The improvement of NUE in crop production is one of the main pillars of this vision [52–54]. Nanotechnology can play an important role in the strengthening of agriculture sustainability, having provided the feasibility of the so-called "smart fertilizer." In other words, nanostructures act as carriers of nutrients and allowed their controlled release.

The design of smart fertilizers strongly influences the nutrient release and the minimization of losses. In field conditions such products are provided to crops via irrigation or sprayed to plant canopies. Through the application of nanotechnologies in agriculture the fertilization will be carried out in different ways. In particular, the nutrient elements will be possibly administered as follows:

i. Delivered as particles or emulsions of nanoscale dimensions: a research body is being developed which aims to clarify whether nanoparticles (e.g., fullerenes, carbon nanotubes, $n\mathrm{TiO_2}$, and $n\mathrm{SiO_2}$) in different growth stages of crops may or may not partially replace traditional fertilizer practices [55, 56].

ii. Encapsulated inside nanostructures designed to allow the controlled release of nutrients (**Figure 1**): to do so the outer shell of nanocapsules is engineered and programmed to open when stimulated by environmental factors or man-induced pulses. Here are some examples of possible control mechanisms [57]:

- Slow release: The capsule releases its payload slowly over a longer period of time so as to synchronize plant assimilation and limit leaching.
- Quick-release: The capsule shell breaks upon contact with a leaf surface.
- Specific release: The nutrient release occurs through a recognition mechanism between a receptor (molecule or functional group) bound to the shell and a target molecule.
- Moisture release: The shell breaks down and releases nutrients in the presence of water.

http://dx.doi.org/10.5772/intechopen.74425

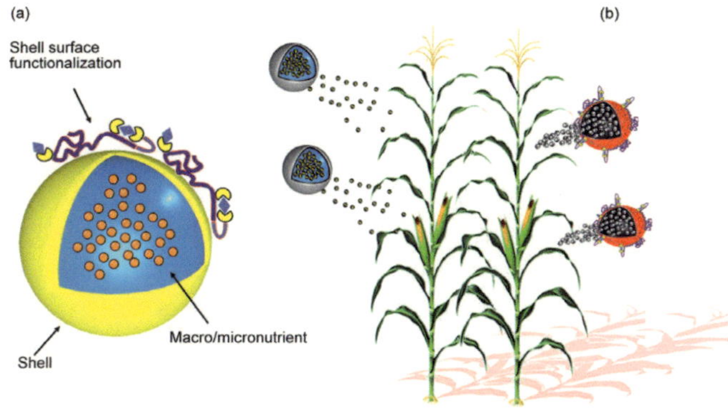

Figure 1. (a) Model of nanocapsule containing macro/microelements. Examples of opening strategies of nanocapsule: (b) release of nutrients as function of time to avoid or limit nutrient losses or designed to occur when a molecular receptor binds to a specific chemical.

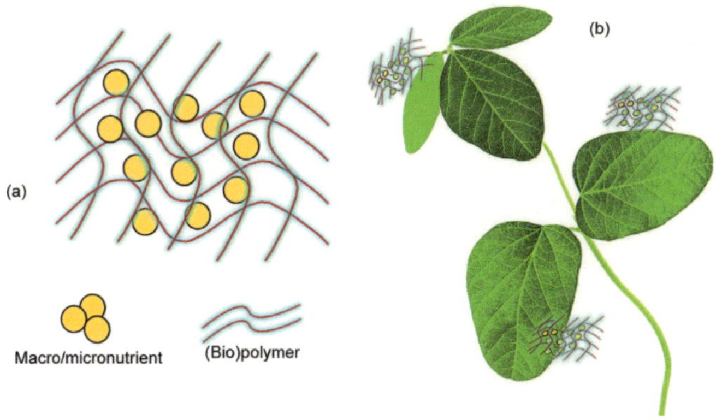

Figure 2. (a) Model of biopolymeric structure containing macro/microelements. (b) Deposition onto the crop leaf after spray treatment.

- pH release: The shell breaks up only in specific alkaline/acidic environment (e.g., within plant tissues or inside a cell).
- Magnetic/ultrasonic pulses: The shell opens in response to a magnetic or ultrasonic pulse emitted by a man-controlled system (precision agriculture).

iii. Delivered in a complex formed by nanocapsules incorporated in a matrix of organic polymers of biological or chemical origin which act as a carrier (**Figure 2**): Both of them provide the expected traits to nanofertilizers. However, natural substances should be preferred as they are easy available, biodegradable and cheaper than the synthetic ones [58]. The properties of the new nanostructure allow a controlled release of nutrients as a function of time or after interactions with the environment. Studies are currently being conducted to test the potential of different materials, such as zeolites [59–61], polyacrylic acid [62] and chitosan [63].

As far as the effectiveness of nanofertilizers is concerned, it must be said that the potential of nanofertilizer application has not been extensively studied yet. However, some successful examples demonstrated that such new formulates significantly improve the efficiency of fertilization [64–70].

The challenge for research is to develop and test carriers that allow the controlled release of nitrogen, following a schedule possibly synchronized with the physiological needs of crops. We are still at a stage where studies on interactions between nanomaterials and biota provide conflicting results. This occurs also for studies on nanofertilizers.

2.2.3. Large-scale use of nanofertilizers

There is no question that nanotechnology is a revolutionary science. However, in several fields of application there are good and bad components to deal with. Referring to nanofertilizers it should be emphasized there are still some uncertainties.

Despite great expectations, both large-scale industrial production of nanofertilizers and their utilization are yet to be realized. This is certainly due to the lack of clear legislative indications. For example, in the European Union, the work to prepare a legislative and regulatory framework is actively under way.

Another controversial point is that, when we look at the recent literature, surprisingly, it can be easily verified that research has neglected macronutrients to focus more in the direction of micronutrients [49–51, 71–73]. This is noteworthy; in fact, although microelements are very important in plant metabolism, crop yield is mainly influenced by N, P and K nutrition.

In conclusion, there are still great expectations that need to be satisfied. In accordance with international and national agencies dealing with sustainable agricultural development and food security (FAO, UNEAP, USEPA, EEA), applied research on nanotechnology in agriculture should be re-oriented according to precise priorities. The development of N and P nanofertilizers is certainly one of such priorities.

2.3. Internet of NanoThings in agriculture

The first definition of precision agriculture (PA) was "an integrated information- and production-based farming system that is designed to increase long term, site-specific and whole farm production efficiency, productivity and profitability while minimizing unintended impacts on wildlife and the environment." It was provided in 1997 by the US House of Representatives [74]. Subsequently the definition narrowed and implemented with the concept of site-specific crop management (SSCM), which is "… a form of PA whereby decisions on resource application and agronomic practices are improved to better match soil and crop requirements as they vary in the field" [75]. This new vision implies that PA is a constantly evolving management strategy, ready to implement—where available—new technologies.

Applications that derive from convergence between Internet of Things (IoT) and nanotechnologies are developing very rapidly in industrial, information and communication technologies, and biomedical sectors. In addition, the future interactions between the Internet and nanodevices introduce a new perspective which has been referred to as the "Internet of NanoThings" (IoNT) [76, 77]. Needless to say, nanotechnology is the new frontier of PA.

2.3.1. Nanobiosensors

Nanobiosensors (NBSs) are analytical devices having at least one dimension no greater than 100 nm. Structured as nanoparticles, nanotubes, nanowires or nanocrystals, NBSs are manufactured for monitoring plant fractions, soil and water in the agroecosystem. By exploiting the physico-chemical properties of nanomaterials, NBSs represent a powerful tool with advanced and improved features compared to existing analytical sensors and biosensors that combine biological element recognition with chemical or physical principles [78]. Biological information is converted by a transducer into a signal yielded by an electronic component. This capability allows the agronomist with an accurate and real-time control of the needs of crops in terms of water and nutrient supply and early symptoms of diseases [79].

A properly designed network of nanosensors would allow the optimization of crop yield and the most efficient agronomic management of factors, such as fertilizers, water, herbicides and pesticides.

Typically, an NBS consists of three components [80, 81]:

i. Biological sensitive probe: a sensing element which interacts with the target (biomolecule) producing a signal proportional to the biomolecule concentration. Some examples of probe/biomolecule interaction are: (*i*) antibody–antigen, (*ii*) nucleic acid interactions (*iii*) enzymatic interactions and (*iv*) cellular interactions (i.e., microorganisms, proteins).

ii. Transducer: a physical component responsible for converting the recognition signal events into a digital signal. The nanomaterial properties suggest managing different kinds of signals such as electrochemical, optical and mass-sensitive signals.

iii. Data recording unit: it consists of an amplifier and signal processor that are responsible for data transferred and storage.

For plant monitoring applications, we therefore deploy a monitoring system comprising a hierarchical arrangement of nano- and microscale network devices (**Figure 3**). The control units manage clusters of nanodevices and the data flow. Data should be directed to gateways which relay the collected data from the nanonetwork to the Internet [82].

Large numbers of nanoscale-sensing devices could be positioned on the plant leaves through suspension in a spray treatment. At this time, this technology is at its very early stage. For its refinement, it will also be necessary to design spraying machines capable of adequately distributing suspensions with nanosensors onto crop canopies.

Nanonetworks for monitoring plant conditions can alert automatically suggesting a more efficient usage of crop inputs (e.g., fertilizers, water, pesticide, etc). Thus, the real time and monitoring of the crop growth lead to accurate and on-time decisions, reduced costs and waste, improved quality of production and above all sustainable agriculture.

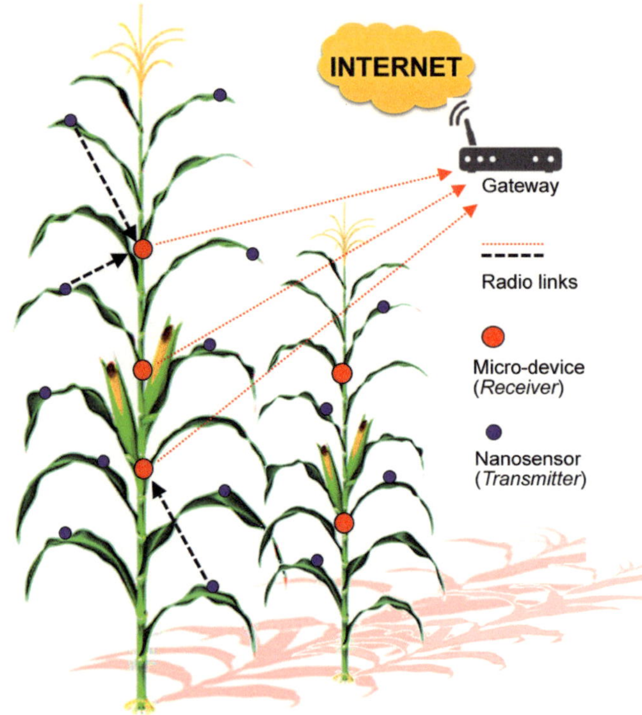

Figure 3. Structure and components of a nanonetwork designed for plant monitoring applications.

Finally, the use of nanobiosensors for high-resolution crop monitoring could be a very useful tool for plant science research. The real-time continuous measurement of plant metabolites and hormones will make a deeper understanding and control of plant biosynthetic pathways in ways not possible.

2.4. Valorization of agricultural residues for production of nanomaterials

There is a growing awareness of the importance of sustainability, in particular bearing in mind the increase of global population [1]. This issue is intimately linked to the implementation of a circular economy based on regeneration of resources. One of the pillars of circular economy is waste reduction.

Organization for Economic Cooperation and Development (OECD) defines agricultural waste as "waste produced as a result of various agricultural operations including manure and other wastes from farms, poultry houses and slaughterhouses; harvest waste; fertilizer run-off from fields; pesticides that enter into water, air or soils; and salt and silt drained from fields" [83].

A meaningful proportion of agri-food production is lost in the form of residues and wastes [84]. For this reason, it will be of the utmost importance to explore innovative technologies capable of providing new opportunities to achieve full sustainability. It is believed that nanotechnology can significantly contribute also in this direction [85]. The development of advanced methods for valorization and the exploitation of agricultural raw materials and wastes are relevant contributions of nanotechnology toward strengthening the basic principles of the circular economy. The following are suggested as illustrative examples of this concept.

2.4.1. Cellulose nanofibers

Cellulose is the most abundant biopolymer available on the Earth, being the main component of plant tissues. The primary occurrence of cellulose is the existing lignocellulosic material in wood which is the most important industrial source of cellulose. Other cellulose-containing materials include agriculture residues, water plants, grasses and other plant substances [86]. It is estimated that 10^{11}–10^{12} tons per year of cellulose are worldwide produced by photosynthesis [87].

In plant tissues micro and macrofibrils represent the construction units of the hierarchical structure of cellulose fibers (**Figure 4**). Microfibrils, in turn, consist of elementary fibrils (nanofibres) which have a diameter comprised in the range 3–35 nm depending on the cellulose source (**Figure 4**) [88].

In recent years, nanocellulose has been attracting much attention as a new bio-based nanomaterial with excellent optical properties, high strength and specific surface area [89, 90]. Nanocellulose can be extracted and chemically modified for a wide range of applications in the field of nanocomposites [91]. Various agricultural crops and residues, such as soy hulls and wheat straw, sugar beet pulp, potato pulp and rutabaga, are already considered as raw materials for new cost-effective methods of nanocellulose production [92–95].

2.4.2. Rice husk-derived Si nanomaterials

FAO's preliminary forecast of global paddy production in 2017 is set at 503.8 million tons (milled basis) [96]. About 25% of this production is rice husk (RH) which is disposed as a by-product of rice milling. The RH is the coating on a grain of rice which has the role to protect the seed during the crop cycle. RHs are mainly composed of lignocellulose (ca. 72–85 wt %) and silica (ca. 15–28 wt %) [97]. Silicon is the second element of importance in the Earth's crust. Grasses assimilate large amounts of Si during their entire life cycle and deposit it into phytoliths as amorphous hydrated silica (SO_2 nH_2O). The Si content in the ash of grasses can reach 50–70% [98].

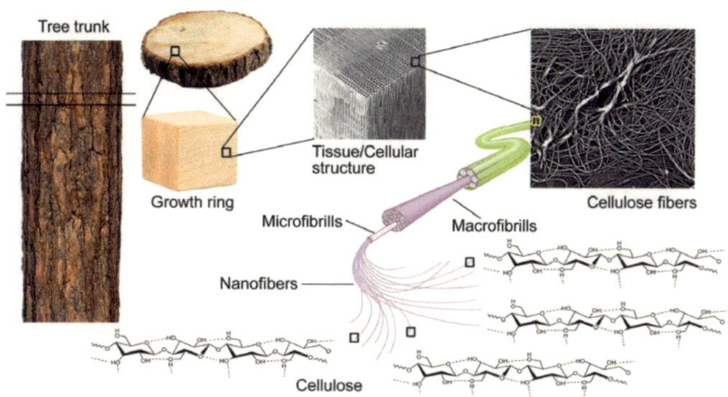

Figure 4. Hierarchical structure of cellulose fibers in wood biomass.

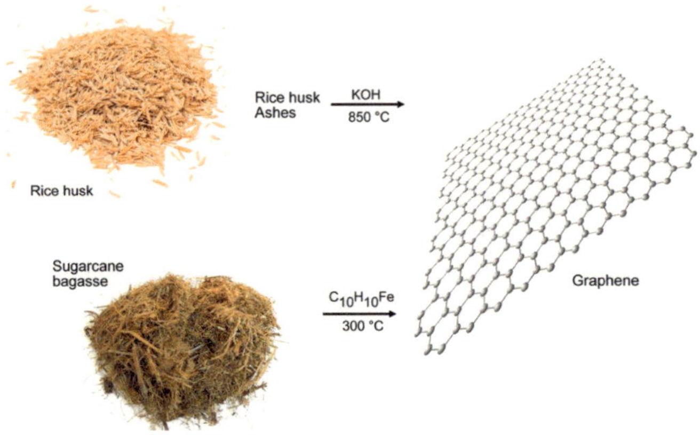

Figure 5. Production of graphene from agricultural wastes.

Silica nanoparticles (*n*Si) have numerous potential applications in drug delivery and biomedicine [99], and in agriculture, as well. According to the principles of green chemistry and among the available agricultural raw materials, RH is considered to be a cost-effective bioprecursor for biosynthesis of *n*Si.

2.4.3. Graphene

Graphene is a material consisting of a monoatomic layer of carbon atoms isolated in 2004 by Novoselov and Geim (University of Manchester, UK), who in 2010 received for that work the Nobel Prize in physics. Graphene has the mechanical strength of the diamond and the flexibility of the plastic and is already used in medicine, electronics, energy, defense and many other sectors. The European Commission, launched in 2013, financed The Graphene Flagship, a 10-year research initiative financed with € 1 billion, which involves more than 140 academic and commercial institutions in 23 countries.

Graphene is currently produced by mechanical and chemical exfoliation of graphite crystals, chemical synthesis and thermal chemical vapor deposition. Considering the large-scale production of graphene, the use of these methods poses several problems due to high process costs and the use of toxic substances. That is why, also in this case, there is considerable interest for the development of alternative, cheaper and environmental-friendly methods.

Recent studies demonstrated that it is possible to use rice husk and sugarcane bagasse to produce graphene in a rapid, scalable and cost-effective manner (**Figure 5**). It is very useful to test other raw agricultural materials to expand the possibility to exploit other wastes or crop byproducts.

3. Conclusions

In this chapter, we have examined some recently developed ideas concerning the possible contributions of nanotechnologies in the primary sector. At this moment, some ideas are, if not completely visionary, strongly projected into the future. Whereas some other hypotheses are very concrete, for some of them, the first experimental data are already available. Thus, in looking ahead to the future, we can be reasonably optimistic.

However, there are a number of concerns linked to the applicative aspects of the use of nanomaterials in agriculture which have to be addressed. How will nanofertilizers (or nanopesticides, nanoherbicides) be handled in field conditions? Which precautionary criteria should be considered? Which equipment or machines will be used? Will these be the same equipment or machines used for bulk materials? What should be the safety conditions for workers? On these aspects, and many others, the authorities will have to define rules. Obviously, on this point, there are great expectations from the industries.

In conclusion, the utilization of nanomaterials in agriculture still needs deep basic knowledge about the fate of nanomaterials in the agro-environment. However, a more mature and, at the same time, a very promising aspect of the interactions between agriculture and nanotechnology are that with regard to the valorization of waste materials. Therefore, it is appropriate to reiterate once again that nanotechnologies are in tumultuous evolution. This means that applications currently under development will soon be overtaken by other ideas that will solve other issues in the field of sustainable agriculture. This principle is nothing but the driving force of the development of knowledge and the strengthening of technology applications.

Acknowledgements

The author is grateful to Regione Friuli Venezia Giulia, General Directorate of Environment, for its financial support to the project "Nanomateriali 2017-2020".

Conflict of interest

The author declares no conflict of interest.

Author details

Luca Marchiol

Address all correspondence to: marchiol@uniud.it

DI4A—Department of Agriculture, Food, Environment and Animal Sciences, University of Udine, Udine, Italy

References

[1] United Nations, Department of Economic and Social Affairs, Population Division. World Population Prospects: The 2017 Revision, Key Findings and Advance Tables. Working Paper No. ESA/P/WP/248. ed. New York: United Nations; 2017

[2] Glenn JC, Florescu E. Millennium project team. 2015-16 state of the future. Journal of Socialomics. 2016;**5**(3):1-6. DOI: 10.4172/2167-0358.1000168

[3] Grierson CS, Barnes SR, Chase MW, Clarke M, Grierson D, Edwards KJ, Jellis GJ, Jones JD, Knapp S, Oldroyd G, Poppy G, Temple P, Williams R, Bastow R. One hundred important questions facing plant science research. New Phytologist. 1921;**2011**:6-12. DOI: 10.1111/j.1469-8137.2011.03859.x

[4] de Rosa MC, Monreal C, Schnitzer M, Walsh R, Sultan Y. Nanotechnology in fertilizers. Nature Nanotechnology. 2010;**5**:91. DOI: 10.1038/nnano.2010.2

[5] de Oliveira JL, Campos EV, Bakshi M, Abhilash PC, Fraceto LF. Application of nanotechnology for the encapsulation of botanical insecticides for sustainable agriculture: Prospects and promises. Biotechnology Advances. 2014;**32**:1550-1561. DOI: 10.1016/j.biotechadv.2014.10.010

[6] Smith P, Bustamante M, Ahammad H, Clark H, Dong H, Elsiddig EA, Haberl H, Harper R, House J, Jafari M, Masera O, Mbow C, Ravindranath NH, Rice CW, Robledo Abad C, Romanovskaya A, Sperling F, Tubiello T. Agriculture, forestry and other land use (AFOLU). In: Edenhofer O, Pichs-Madruga R, Sokona Y, Farahani E, Kadner S, Seyboth K, Adler A, Baum I, Brunner S, Eickemeier P, Kriemann B, Savolainen J, Schlömer S, von Stechow C, Zwickel T, Minx JC, editors. Climate Change 2014: Mitigation of Climate Change. Contribution of Working Group III to the Fifth Assessment Report of the Intergovernmental Panel on Climate Change. Cambridge, United Kingdom and New York, NY, USA: Cambridge University Press; 2014. p. 811-922

[7] EUROSTAT. Total waste generation in the EU-27 by economic activity in 2008 [Internet]. Available from: http://ec.europa.eu/eurostat/statistics-explained/index.php/Waste_statistics [Accessed: October 15, 2017]

[8] EC, Communication from the Commission to the European Parliament, the Council, the European Economic and Social Committee and the Committee of the Regions. A European strategy for Key Enabling Technologies - A bridge to growth and jobs [Internet]. Available from: http://eur-lex.europa.eu/LexUriServ/LexUriServ.do?uri=COM:2012:0341:FIN:EN:PDF [Accessed: October 15, 2017]

[9] Kagan CR. At the nexus of food security and safety: Opportunities for nanoscience and nanotechnology. ACS Nano. 2016;**10**:2985-2986. DOI: 10.1021/acsnano.6b01483

[10] Nazim M, Ahmad M. A bibliometric analysis on nanotechnology research. Annals of Library and Information Studies. 2008;**55**:292-299

[11] Mani PK, Mondal S. Agri-nanotechniques for plant availability. In: Kole C, Sakthi Kumar D, Khodakovskaya MV, editors. Plant Nanotechnology. Principles and Practices. Switzerland: Springer International Publishing; 2016. pp. 263-303. DOI: 10.1007/978-3-319-42154-4_11

[12] Wang P, Lombi E, Zhao F, Kopittke PM. Nanotechnology: A new opportunity in plant sciences. Trends Plant Science. 2016;**218**:699-712. DOI: 10.1016/j.tplants.2016.04.005

[13] JRC report. Proceedings of a workshop on "Nanotechnology for the agricultural sector: from research to the field [Internet]. 2014. Available from: https://ec.europa.eu/jrc/sites/jrcsh/files/ipts_jrc_89736_(online)__final.pdf

[14] RIKILT and JRC. Inventory of nanotechnology applications in the agricultural, feed and food sector. EFSA supporting publication. 2014:125

[15] Gogos A, Knauer K, Bucheli TD. Nanomaterials in plant protection and fertilization: Current state, foreseen applications, and research priorities. Journal of Agriculture and Food Chemistry. 2012;**60**:9781-9792. DOI: 10.1021/jf302154y

[16] Chen H, Yada R. Nanotechnologies in agriculture: New tools for sustainable development. Trends in Food Science & Technology. 2011;**22**:585-594. DOI: 10.1016/j.tifs.2011.09.004

[17] Wu A, Song Y, van Oosterom EJ, Hammer GL. Connecting biochemical photosynthesis models with crop models to support crop improvement. Frontiers in Plant Science. 2016;**7**: 1518. DOI: 10.3389/fpls.2016.01518

[18] FAO. World agriculture towards 2030/2050: the 2012 revision [Internet]. 11/06/212. Available from: http://www.fao.org/docrep/016/ap106e/ap106e.pdf [Accessed: October 24, 2017]

[19] Long SP, Zhu X-G, Naidu SL, Ort DR. Can improvement in photosynthesis increase crop yields? Plant Cell Environment. 2006;**29**:315-330. DOI: 10.1111/j.1365-3040.2005.01493.x

[20] Spreitzer RJ, Salvucci ME. Rubisco: Structure, regulatory interactions, and possibilities for a better enzyme. Annual Reviews in Plant Biology. 2002;**53**:449-475. DOI: 10.1146/annurev.arplant.53.100301.135233

[21] McGrath JM, Long SP. Can the cyanobacterial carbon-concentrating mechanism increase photosynthesis in crop species? A theoretical analysis. New Phytologist. 1644;**2014**:2247-2261. DOI: 10.1104/pp.113.232611

[22] Tan DKY, Amthor JS. Bioenergy. In: Dubinsky Z, editor. Photosynthesis. Intech; 2013. pp. 299-330. DOI: 10.5772/55317

[23] Zhu X-G, Long SP, Ort DR. Improving photosynthetic efficiency for greater yield. Annual Reviews in Plant Biology. 2010;**61**:235-261. DOI: 10.1146/annurev-arplant-042809-112206

[24] Hibberd JM, Sheehy JE, Langdale JA. Using C4 photosynthesis to increase the yield of rice - rationale and feasibility. Current Opinion in Plant Biology. 2008;**112**:228-231. DOI: 10.1016/j.pbi.2007.11.002

[25] Amthor J. Improving photosynthesis and yield potential. In: Ranalli P, editor. Improvement of Crop Plants for Industrial End Uses. Dordrecht: Springer; 2007. pp. 27-58. DOI: 10.1007/978-1-4020-5486-0_2

[26] Melis A. Solar energy conversion efficiencies in photosynthesis: Minimizing the chlorophyll antennae to maximize efficiency. Plant Science. 1774;**2009**:272-280. DOI: 10.1016/j.plantsci.2009.06.005

[27] Wang Q, Zhang QD, Zhu XG, Lu CM, Kuang TY, Li CQ. PSII photochemistry and xanthophyll cycle in two superhigh-yield rice hybrids, Liangyoupeijiu and Hua-an 3 during photoinhibition and subsequent restoration. Acta Botanica Sinica. 2002;**4411**:1297-1303

[28] Evans JR. Improving photosynthesis. Plant Physiology. 2013;**162**:1780-1793. DOI: 10.1104/pp.113.219006

[29] Hong F, Zhou J, Liu C, Yang F, Wu C, Zheng L, Yang P. Effect of nano-TiO_2 on photochemical reaction of chloroplasts of spinach. Biology Trace Elements Research. 2005a;**105** (1-3):269-279. DOI: 10.1385/BTER:105:1-3:269

[30] Hong FS, Yang F, Ma ZN, Zhou J, Liu C, Wu C, Yang P. Influences of nano-TiO_2 on the chloroplast ageing of spinach under light. Biology Trace Elements Research. 2005b;**1043**: 249-260. DOI: 10.1385/BTER:104:3:249

[31] Gao FQ, Hong FS, Liu C, Zheng L, Su MY. Mechanism of nano-anatase TiO_2 on promoting photosynthetic carbon reaction of spinach: Inducing complex of Rubisco–Rubisco activase. Biology Trace Elements Research. 2006;**111**(1-3):239-253. DOI: 10.1385/BTER:111:1:239

[32] Gao FQ, Liu C, Qu CX, Zheng L, Yang F, Su MG, Hong FH. Was improvement of spinach growth by nano-TiO_2 treatment related to the changes of rubisco activase? Biometals. 2008;**21**:211-217. DOI: 10.1007/s10534-007-9110-y

[33] Ma L, Liu C, Qu C, Yin S, Liu J, Gao F, Hong F. Rubisco activase mRNA expression in spinach: Modulation by nanoanatase treatment. Biology Trace Elements Research. 2008; **1222**:168-178. DOI: 10.1007/s12011-007-8069-4

[34] Qi M, Liu Y, Li T. Nano-TiO_2 improve the photosynthesis of tomato leaves under mild heat stress. Biology Trace Elements Research. 2013;**156**(1-3):323-328. DOI: 10.1007/s12011-013-9833-2

[35] Lei Z, Mingyu S, Chao L, Liang C, Hao H, Xiao W, Xiaoqing L, Fan Y, Fengqing G, Fashui H. Effects of nanoanatase TiO_2 on photosynthesis of spinach chloroplasts under different light illumination. Biology Trace Elements Research. 2007;**1191**:68-76. DOI: 10.1007/s12011-007-0047-3

[36] Giraldo JP, Landry MP, Faltermeier SM, McNicholas TP, Iverson MN, Boghossian AA, Reuel NF, Hilmer AJ, Sen F, Brew J, Strano MS. Plant nanobionics approach to augment

photosynthesis and biochemical sensing. Nature Materials. 2014;**13**:400-408. DOI: 10.1038/nmat3890

[37] Noji T, Kamidaki C, Kawakami K, Shen JR, Kajino T, Fukushima Y, Sekitoh T, Itoh S. Photosynthetic oxygen evolution in mesoporous silica material: Adsorption of photosystem II reaction center complex into 23 nm nanopores in SBA. Langmuir. 2011;**272**:705-713. DOI: 10.1021/la1032916

[38] Singha MR, Bell AT. Design of an artificial photosynthetic system for production of alcohols in high concentration from CO_2. Energy & Environmental Science. 2016;**9**:193-199. DOI: 10.1039/C5EE02783G

[39] Marshner H. Mineral Nutrition of Higher Plants. 3rd ed. London: Elsevier/Academic Press; 2012. p. 672. DOI: 10.1016/B978-0-12-384905-2.00030-3

[40] Stewart WM, Dibb DW, Johnston AE, Smyth TJ. The contribution of commercial fertilizer nutrients to food production. Agronomy Journal. 2005;**7**(1):1-6. DOI: 10.2134/agronj2005.0001

[41] Baligar VC, Fageria NK. Nutrient use efficiency in plants: An overview. In: Rakshit A, Singh HB, Sen A, editors. Nutrient Use Efficiency: From Basics to Advances. New Delhi: Springer; 2015. pp. 1-14. DOI: 10.1007/978-81-322-2169-2_1

[42] Eriksen A, Kjeldby M. A comparative study of urea hydrolysis rate and ammonia volatilization from urea and urea calcium nitrate. Fertilizer Research. 1987;**11**(1):9-24. DOI: 10.1007/BF01049561

[43] Bouwman L, Goldewijk KK, Van Der Hoek KW, Beusen AHW, Van Vuuren DP, Willems J, Rufino MC, Stehfesta E. Exploring global changes in nitrogen and phosphorus cycles in agriculture induced by livestock production over the 1900–2050 period. Proceedings National Academy of Science USA. 2013;**110**(52):20882-20887. DOI: 10.1073/pnas.1012878108

[44] Welch RM, Shuman L. Micronutrient nutrition of plants. Critical Reviews in Plant Sciences. 1995;**14**(1):49-82. DOI: 10.1080/07352689509701922

[45] Alloway BJ. Heavy metals and metalloids as micronutrients for plants and animals. In: Alloway B, editor. Heavy Metals in Soils. Environmental Pollution. Vol. 2013, 22. Dordrecht: Springer; 2013. pp. 195-209. DOI: 10.1007/978-94-007-4470-7_7

[46] Mortved JJ. Micronutrient fertilizers and fertilization practices. Nutrient Cycling in Agroecosystems. 1985;**7**:1,221-1,235. DOI: 10.1007/BF01049002

[47] Monreal CM, DeRosa M, Mallubhotla SC, Bindraban PS, Dimkpa C. Nanotechnologies for increasing the crop use efficiency of fertilizer-micronutrients. Biology Fertility Soils. 2016;**52**:423-437. DOI: 10.1007/s00374-015-1073-5

[48] Page V, Feller U. Heavy metals in crop plants: Transport and redistribution processes on the whole plant level. Agronomy. 2015;**5**:447-463. DOI: 10.3390/agronomy5030447

[49] Sutton MA, Bleeker A, Howard CM, Erisman JW, Abrol YP, Bekunda M, Datta A, Davidson E, de Vries W, Oenema O, Zhang FS. Our nutrient world. The challenge to produce more food & energy with less pollution. Key messages for Rio+20. Published by the Centre for Ecology & Hydrology on Behalf of the Global Partnership on Nutrient Management (GPNM) and the International Nitrogen Initiative (INI), 2012

[50] Chhipa H. Nanofertilizers and nanopesticides for agriculture. Environmental Chemistry Letters. 2016;**15**(1):15-22. DOI: 10.1007/s10311-016-0600-4

[51] Zuverza-Mena N, Martínez-Fernández D, Du W, Hernandez-Viezcas JA, Bonilla-Bird N, López-Moreno ML, Komárek M, Peralta-Videa JR, Gardea-Torresdey JL. Exposure of engineered nanomaterials to plants: Insights into the physiological and biochemical responses-a review. Plant Physiology Biochemistry. 2017;**110**:236-264. DOI: 10.1016/j.plaphy.2016.05.037

[52] Manjunatha SB, Biradar DP, Aladakatti YR. Nanotechnology and its applications in agriculture: A review. Journal Farm Science. 2016;**29**:1-13

[53] Ramos Campos EV, de Oliveira JL, Fernandes Fraceto L, Singh B: Polysaccharides as safer release systems for agrochemicals. Agronomy for Sustainable Development. 2015;**35**:47-66. DOI 10.1007/s13593-014-0263-0

[54] Morales-Díaz AB, Ortega-Ortíz H, Juárez-Maldonado A, Cadenas-Pliego G, González-Morales S, Adalberto Benavides-Mendoza A. Application of nanoelements in plant nutrition and its impact in ecosystems. Advances in Natural Sciences: Nanoscience and Nanotechnology. 2017;**8**:1-13. DOI: 10.1088/2043-6254/8/1/013001

[55] Millan G, Agosto F, Vazquez M. Use of clinoptilolite as a carrier for nitrogen fertilizers in soils of the Pampean regions of Argentina. Ciencia Investigación Agraria. 2008;**35**:293-302

[56] Perrin TS, Drost DT, Boettinger JL, Norton JM. Ammonium-loaded clinoptilolite: A slow-release nitrogen fertilizer for sweet corn. Journal Plant Nutrition. 1998;**21**:515-530. DOI: 10.1080/01904169809365421

[57] Aouada FA, de Moura MR. Nanotechnology applied in agriculture: Controlled release of agrochemicals. In: Rai M, Ribeiro C, Mattoso L, Duran N, editors. Nanotechnologies in Food and Agriculture. Springer, Cham. 2015. p. 103-118. DOI: 10.1007/978-3-319-14024-7_5

[58] Corradini E, de Moura MR, Mattoso LHC. A preliminary study of the incorparation of NPK fertilizer into chitosan nanoparticles. eXPRESS Polymer Letters. 2010;**4**(8):509–515. DOI: expresspolymlett.2010.64

[59] Ditta A, Arshad M. Applications and perspectives of using nanomaterials for sustainable plant nutrition. Nanotechnology Reviews. 2016;**5**(2):1-22. DOI: 10.1515/ntrev-2015-0060

[60] Dimkpa CO, Bindraban PS. Nanofertilizers: New products for the industry? Journal of Agriculture and Food Chemistry. . DOI: 10.1021/acs.jafc.7b02150

[61] Servin A, Elmer W, Mukherjee A, De la Torre-Roche R, Hamdi H, White JC, Bindraban P, Dimkpa C. A review of the use of engineered nanomaterials to suppress plant disease and

enhance crop yield. Journal of Nanoparticle Research. 2015;**17**:92. DOI 10.1007/s11051-015-2907-7

[62] US House of Representatives (1997): Section 2.1.A Precision Agriculture. Online source: http://www.gpo.gov/fdsys/pkg/BILLS-105hr725ih/html/BILLS-105hr725ih.htm

[63] Robert PC, Rust RH, WEA L, editors. Site-Specific Management for Agricultural Systems. Madison, WI: ASA-CSSA-SSSA; 1995. 933 p

[64] Nayyar A, Puri V, Le D-N. Internet of nano things (IoNT): Next evolutionary step in nanotechnology. Nanoscience and Nanotechnology. 2017;**7**(1):4-8. DOI: 10.5923/j.nn.20170701.02

[65] Ezz El-Din H, Manjaiah DH. Internet of nano things and industrial internet of things. In: Acharjya D, Geetha M, editors. Internet of Things: Novel Advances and Envisioned Applications. Studies in Big Data. Cham: Springer; 2017. pp. 109-123. DOI: 10.1007/978-3-319-53472-5_5

[66] Fraceto LF, Grillo R, de Medeiros GA, Scognamiglio V, Rea G, Bartolucci C. Nanotechnology in agriculture: Which innovation potential does it have? Frontiers in Environmental Science. 2016;**4**:20. DOI: 10.3389/fenvs.2016.00020

[67] Mousavi SR, Rezaei M. Nanotechnology in agriculture and food production. Journal of Applied Environmental and Biological Sciences. 2011;**1**:10,414-10,419

[68] Prasad R, Bhattacharyya A, Nguyen QD. Nanotechnology in sustainable agriculture: Recent developments, challenges, and perspectives. Frontiers in Microbiology. 2017;**8**: 1014. DOI: 10.3389/fmicb.2017.01014

[69] Mufamadi MS, Sekhejane PR. Nanomaterial-based biosensors in agriculture application and accessibility in rural smallholding farms: Food security. In: Prasad R, Kumar M, Kumar V, editors. Nanotechnology. Singapore: Springer; 2017. pp. 263-278. DOI: 10.1007/978-981-10-4573-8_12

[70] Afsharinejad A, Davy A, Jennings B. Dynamic channel allocation in electromagnetic nanonetworks for high resolution monitoring of plants. Nano Communication Networks. 2016;**7**:2-16. DOI: 10.1016/j.nancom.2015.01.001

[71] OECD. Glossary of Statistical Terms. Paris; 2009

[72] Ashworth GS, Azevedo P, editors. Agricultural Wastes. Nova Science Publishers, Inc; 2009. 306 p. ISBN: 978-1-60741-305-9

[73] Parisi C, Vigani M, Rodríguez-Cerezo E. Proceedings of a Workshop on Nanotechnology for the Agricultural Sector: From Research to the Field. EUR – Scientific and Technical Research Series. 2014. 40 p. DOI 10.2791/80497

[74] Klemm D, Schmauder H-P, Heinze T. Cellulose. In: de Baets S,Vandamme E, Steinbüchel A, editors. Biopolymers. vol. 6. Polysaccharides II: Polysaccharides from Eukaryotes. Wiley-VCH Weinheim, 2004. p. 275-319

[75] Nechyporchuk O, Naceur Belgacem M, Bras J. Production of cellulose nanofibrils: A review of recent advances. Industrial Crops and Products. 2016;**93**:2-25. DOI: 10.1016/j.indcrop.2016.02.016

[76] Bagal-Kestwal DR, Kestwal RM, Chiang BH. Bio-based nanomaterials and their bionanocomposites. In: Visakh PM, Martínez Morlanes MJ, editors. Nanomaterials and Nanocomposites: Zero- to Three-Dimensional Materials and their Composites. 1st ed. Wiley; 2016. pp. 255-329. DOI: 10.1002/9783527683772.ch8

[77] Kallel F, Bettaieb F, Khiari R, GarcíaA, Bras J, Chaabouni SE. Isolation and structural characterization of cellulose nanocrystals extracted from garlic straw residues. Industrial Crops and Products. 2016;**87**:287-296. DOI: 10.1016/j.indcrop.2016.04.060

[78] Khiari R. Valorization of agricultural residues for cellulose nanofibrils production and their use in nanocomposite manufacturing. International Journal of Polymer Science. 2017;**9**:1-10. DOI: 10.1155/2017/6361245

[79] Mariano M, El Kissi N, Dufresne A. Cellulose nanocrystals and related nanocomposites: Review of some properties and challenges. Journal of Polymer Science, Part B: Polymer Physics. 2014;**52**(12):791-806. DOI: 10.1002/polb.23490

[80] Espinosa E, Tarrés Q, Delgado-Aguilar M, González I, Mutjé P, Rodríguez A. Suitability of wheat straw semichemical pulp for the fabrication of lignocellulosic nanofibres and their application to papermaking slurries. Cellulose. 2015;**23**(1):837-852. DOI: 10.1007/s10570-015-0807-8

[81] Habibi Y, Vignon MR. Optimization of cellouronic acid synthesis by TEMPO-mediated oxidation of cellulose III from sugar beet pulp. Cellulose. 2008;**15**(1):177-185. DOI: 10.1007/s10570-007-9179-z

[82] Dufresne A, Dupeyre D, Vignon MR. Cellulose microfibrils from potato tuber cells: Processing and characterization of starch-cellulose microfibril composites. Journal of Applied Polymer Science. 2000;**76**:14,2080-14,2092. DOI: 10.1002/(SICI)1097-4628(20000628)76:143.0.CO;2-U

[83] Bruce DM, Hobson RN, Farrent JW, Hepworth DG. High-performance composites from low-cost plant primary cell walls. Composites Part A: Applied Science and Manufacturing. 2005;**36**(11):1486-1493. DOI: 10.1016/j.compositesa.2005.03.008

[84] FAO. Rice market Monitor. 2017. Available from: http://www.fao.org/economic/RMM [Accessed December 1, 2017]

[85] Chen H, Wang W, Martin JC, Oliphant AJ, Doerr PA, Xu JF, DeBorn KM, Chen C, Sun L. Extraction of lignocelluloses and synthesis of porous silica nanoparticles from rice husks: A comprehensive utilization of rice husk biomass. ACS Sustainable Chemistry & Engineering. 2013;**1**:254-259. DOI: 10.1021/sc300115r

[86] Lanning FC, Ponnaiya BWX, Crumpton CF. The chemical nature of silica in plants. Plant Physiology. 1958;**33**:339-343

[87] Wang Y, Zhao Q, Han N, Bai L, Li J, Liu J, Che E, Hu L, Zhang Q, Jiang T, Wang S. Mesoporous silica nanoparticles in drug delivery and biomedical applications. Nanomedicine: Nanotechnology, Biology and Medicine. 2015;**11**(2):313-327. DOI: 10.1016/j.nano.2014.09.014

[88] Athinarayanan J, Periasamy VS, Alhazmi M, Alatiah KA, Alshatwi AA. Synthesis of biogenic silica nanoparticles from rice husks for biomedical applications. Ceramics International. 2015;**41**:275-281. DOI: 10.1016/j.ceramint.2014.08.069

[89] Wanyika H, Gatebe E, Kioni P, Tang Z, Gao Y. Mesoporous silica nanoparticles carrier for urea: Potential applications in agrochemical delivery systems. Journal of Nanoscience and Nanotechnology. 2012;**12**:2221-2222. DOI: 10.1166/jnn.2012.5801

[90] Anastas PT, Eghbali N. Green chemistry: Theory and practice. Chemical Society Reviews. 2010;**39**:301-312. DOI: 10.1039/B918763B

[91] Shen Y. Rice husk silica derived nanomaterials for sustainable applications. Renewable and Sustainable Energy Reviews. 2017;**80**:453-466. DOI: 10.1016/j.rser.2017.05.115

[92] Sankar S, Sharma SK, Kaur N, Lee B, Kim DY, Lee S, Jung H. Biogenerated silica nanoparticles synthesized from sticky, red, and brown rice husk ashes by a chemical method. Ceramics International. 2016;**42**:4875-4885. DOI: 10.1016/j.ceramint.2015.11.172

[93] Novoselov KS, Geim AK, Morozov SV, Jiang D, Zhang Y, Dubonos SV, Grigorieva IV, Firsov AA. Electric field effect in atomically thin carbon films. Science. 2004;**306**(5696):666-669. DOI: 10.1126/science.1102896

[94] The home of graphene. http://www.graphene.manchester.ac.uk/ [Accessed: December 3, 2017]

[95] Graphene Flagship. Available from http://graphene-flagship.eu/ [Accessed: December 3, 2017]

[96] Bhuyan SA, Uddin N, Islam M, Bipasha FA, Hossain SS. Synthesis of graphene. International Nano Letters. 2016;**6**:65-83. DOI: 10.1007/s40089-015-0176-1

[97] Muramatsu H, Kim YA, Yang K-S, Cruz-Silva R, Toda I, Yamada T, Terrones M, Endo M, Hayashi T, Saitoh H. Rice husk-derived graphene with nano-sized domains and clean edges. Small. 2014;**10**(14):2766-2770. DOI: 10.1002/smll.201400017

[98] Somanathan T, Prasad K, Ostrikov K, Saravanan A, Krishna MV. Graphene oxide synthesis from agro waste. Nanomaterials. 2015;**5**:826-834. DOI: 10.3390/nano5020826

[99] Baker S, Volovab T, Prudnikovac SV, Satishd S, Prasad NMN. Nanoagroparticles emerging trends and future prospect in modern agriculture system. Environmental Toxicology and Pharmacology. 2017;**53**:10-17. DOI: 10.1016/j.etap.2017.04.012

Chapter 8

Risks, Uncertainties, and Ethics of Nanotechnology in Agriculture

Phenny Mwaanga

Additional information is available at the end of the chapter

http://dx.doi.org/10.5772/intechopen.76590

Abstract

The use of agrochemicals, though has increased the agricultural productivity, has severely adversely affected soil and aquatic systems with associated flora and fauna and also the health of the farmers and society consuming the chemically grown food. Therefore, the advent of nano-agrochemicals, such as nanopesticides, nanofertilizers and nanosensors, designed to increase solubility, enhances bioavailability and promotes targeted delivery, and their controlled release will have immense potential benefits that include efficient dosage of fertilizers, improved vector and pest management, reduced chemical pollution and ultimately increased agricultural productivity. However, many questions remain unresolved on the toxicology and safety of these systems to human and ecosystems health. Risk assessment of this technology lags far behind its application. This chapter will therefore discuss the nano-ecotoxicology and risks, uncertainties, and ethical concerns of use of nanotechnology in agriculture. Furthermore, the current levels of public awareness and perception of nanotechnology will be discussed.

Keywords: nanotechnology, agriculture, nano-ecotoxicology, health risks, uncertainties, public perception, awareness

1. Introduction

The world including the developing world has seen an extraordinary growth in agricultural food crop productivity in the last 5 decades [1]. Although there are still a few reported shortages of food and incidences of hunger, particularly in few low-income countries, the reason for such food shortages is partly attributed to poor or little application of science and technology in agriculture [2]. But overall, according to some available data [1], despite the world population having more than doubled during the last 5 decades, the production of cereal crops tripled during this same period, with only a 30% increase in the cultivated land area. Thus, if the food currently produced was to be equitably distributed, there would be no person going hungry as there is more food produced than the world population needs. This increased agricultural productivity is largely attributed to the use of agrochemicals, fertilizers, and chemical pesticides [1, 3–7].

The use of fertilizers and pesticides is considered as the panacea for improved crop production [4–6]. Despite the high cost of fertilizers and pesticides, farmers are always availed with these inputs as governments in many countries provide subsidies, with a sole purpose of increasing food security. The optimal benefits from the use of these agrochemicals to a larger extent are realized, except in a few cases, particularly from some low-income countries, where farmers lack technical information such as optimum doses, correct method and right time of application. It should be understood that the requirement of fertilizers and pesticides for crops differs according to soil types and meteorology [7], and where this understanding is not followed, the increase in the use of fertilizers and pesticides may not necessarily correspond to the increased crop productivity [2], and this may be exacerbated by inability to embrace science and technology and the sole dependency on rain-fed agriculture, particularly in sub-Saharan Africa (SSA) region, where rainfall is usually erratic [5]. Elsewhere [7] huge quantities of these agrochemicals are applied to the fields under the adage that "more is better" without necessarily taking into account the soil, meteorology and other factors.

While fertilizers help in adding the nutrients into the soils required for optimal crop growth, excessive and repeated use of fertilizers can result into serious pollution. In some cases, particularly in low-income countries, the application of chemical fertilizers is done without regard to appropriate doses [2]. World over, there are many places where chemical fertilizers have left a legacy of serious pollution particularly for aquatic systems. For instance in Thailand, in Nakhon Pathom Province, in a survey conducted by some Thai scientists to determine the nitrate levels, it was found that 46% of tested ground water had elevated levels of nitrate above the WHO drinking water safety limit of 50 mg/L NO_3^- and this was attributed to agricultural activities [3]. Similarly, in rural settings of Andhra Pradesh, India, as much as 50–70% of the water resources are polluted due to contamination from agricultural activities [8]. To a large extent, this contamination results from applied synthetic nitrogen fertilizers that are unutilized by crops, which in some cases may be as much as 95% [8]. Water pollution, both for surface and ground water, from chemical fertilizers has been reported to affect many countries, including those from the European Union and other developed countries [9], and therefore, mitigation efforts require not only an integrated approach, but also a paradigm shift.

Similarly, pesticide pollution from agricultural activities is quite extensive. It is estimated that about 2.5 million tonnes of chemical pesticides are used on agricultural crops each year [10]. The repeated use of pesticides unfortunately increases pest resistance, and it is this resistance that leads to progressive increase in the amount of the applied pesticides, and sometimes this can impact the food quality. The overuse of pesticides, particularly in low-income countries, due to low literacy levels, can increase agricultural cost and generate considerable waste and pollution, which ultimately adversely affects human health and the environment. The extent of this pollution is evidenced by the pesticide residues that have been detected and quantified in a variety of common foods and beverages, including for instance, animal products, water, wine, fruits, vegetables and animal feeds [7, 11]. Many chemical pesticides have been associated with human health and environmental adverse effects [11]. For instance, some specific adverse human health effects associated with chemical pesticides include among others, dermatological, neurological, teratogenic, clastogenic, carcinogenic, respiratory, reproductive and endocrine effects [5, 7, 11]. The incidences of adverse human health effects from chemical pesticides are disproportionally much more prevalent in developing than in developed countries, where the majority of users have low literacy levels [5]. Ironically, while pesticides have drastically reduced agricultural crop losses, both for preharvest and postharvest, their residue levels in food stuffs, soils, flora and fauna, and water has escalated, thereby posing great risks to the farmers and consumers, including some organisms that are far removed from agricultural sites.

The continued use of these agrochemicals has led to increased levels of pollution and contamination of both aquatic and terrestrial systems with attendant adverse effects on biota. Solving these problems requires an integrated approach and a complete paradigm shift. Thus, not only the development of new and less toxic agrochemicals is necessary and urgent, but also safe, smart and efficient application methods are essential for preventing accumulation and ultimately the adverse effects on the environment. In this vain, nanotechnology offers great promise and can be used as an innovative tool for delivering agrochemicals smartly and safely [10]. Therefore, the advent of nano-agrochemicals, such as nanopesticides, nanofertilizers and nanosensors, designed to increase solubility, enhance bioavailability, promote targeted delivery and controlled release will have immense potential benefits that include efficient dosage of fertilizers, improved vector and pest management, reduced chemical pollution and ultimately increased agricultural productivity. However, many questions remain unresolved on the toxicology and safety of these systems to human and ecosystems health. Currently, the development of this technology (nanotechnology) for use in agriculture has outpaced the risk assessment, thereby posing great challenges on its acceptability by the general public and ultimately may negatively impact on the potential investment by the agricultural industry. This chapter will, therefore, contain discussions on the nano-ecotoxicology and risks, uncertainties and ethical concerns of use of nanotechnology in agriculture. Furthermore, this chapter will review the current levels of public awareness and perception of nanotechnology in agriculture.

2. Nano-ecotoxicology and risks

2.1. Nanotechnology and nanoparticles

A review of literature reveals a multitude of definitions for nanoparticles (NPs) or nanomaterials. In this chapter, a nanoparticle is defined as any intentionally produced particle that has a characteristic dimension from 1 to 100 nm and has properties that are different from that of non-nanoscale particle with the same chemical composition [12, 13]. It is well known that nanoparticles (NPs) may be naturally occurring or intentionally produced. Naturally occurring NPs result from natural processes [14] and are as old as nature, while intentionally produced NPs are often referred to as engineered nanoparticles (ENPs) and are manufactured either by the top-down approach or bottom-up approach [15]. Nanotechnology, therefore, can be defined as the design, characterization, production and application of structures, devices and systems by controlling shape and size at a nanometer scale [14]. Thus, the incorporation of nanoparticles/nanomaterials in the production of goods for application in various fields such as medicine, information and communication technology, engineering, environmental remediation, among others falls within the wider domain of nanotechnology.

NPs, due to their small sizes, have increased relative surface area and the quantum effects that have been observed to dominate the behavior of matter at the nanoscale. It is these factors that can change or enhance properties, such as strength, chemical, biological, electronic, rheological, magnetic, optical (photon), mechanical and structural, and reactivity characteristics [12, 16]. Some researchers [12] have argued that the occurrence of the novel size-dependent properties, rather than particle size, should be the primary criterion when considering the regulation of NPs for environmental, health and safety reasons. Thus, the fact that the particles or materials fit within the definition of a nano may not necessarily exhibit the "nanoness," that is, the occurrence of the novel size-dependent properties, and the size at which this nanoness is observed may be different for different materials. There are several implications of this observation in the regulation of NPs for human and environmental impacts. The first one is that the risk assessment of NPs will have to take into account the size for each material at which the nanoness is observed. Secondly, any material at nanoness, including biological NPs, can potentially cause adverse effects. Thirdly, there is a need for proper NP characterization prior to risk assessment if the risk assessment data are to be comparable. Finally, because at nanoness there is a dramatic increase in surface reactivity, most NPs tend to have increased solubility and this can pose a challenge in delineating toxic effects due to NPs from that due to dissolved ions. This last point is particularly applicable for inorganic-based NPs.

2.2. Nano-ecotoxicology and risks

The "nanoness" properties of NPs, the surface structure and reactivity are responsible for processes such as dissolution, redox reactions and the generation of reactive oxygen species (ROS) [14, 15, 17]. These are the properties that can lead to biological/toxicological effects that would not be produced by bulk particles of the same chemical composition.

Whether or not a given nanoparticle/nanomaterial will induce ecotoxicological effects on an organism upon contact, ingestion or inhalation will depend to a larger extent on its "nanoness." As argued elsewhere [12], most ENPs are likely to be of human or environmental health concern owing to their unique properties only when they have diameters of 30 nm or less. In assessing the potential adverse effects and hence the risks NPs pose to human health, a number of toxicity tests have been conducted in various media and using a variety of organisms. Literature is replete with studies that have been conducted both in vitro and in vivo, although with some conflicting results, even with the same organism for the same type of NPs. One of the possible explanations for the conflicting results would be either due to nonadherence to NP characterization requirements prior to toxicity testing or lack of NP risk assessment guidelines or both. In order to reduce such dichotomy in toxicity results, researchers have tried to understand the best dose metrics that would define the toxicology of the NPs. For instance, researchers have investigated whether a given NP type induces its toxicity through its particle charge, number concentration, mass concentration, total surface area or simply by size. Knowledge of dose metrics responsible for toxic effects as stated by some researchers [18, 19] can have a number of advantages that include easy of adaptation of the risk assessment data into the regulatory framework that ensure the safe use of such NPs, particularly in agriculture, easy of comparison of study results and hence enable regulators to formulate health-based limit values for each metric. And finally, this can also help risk assessors to compare and combine exposure and hazard information and conclude on the likelihood of health risks of each NP type.

In nano-ecotoxicology and risk assessment, various types of NPs such as carbon-, inorganic- and organic-based NPs have been investigated. This is because these types of NPs have found wide application in various fields, particularly in agriculture. Although the application of NPs in agriculture is still largely in the developing stage, there is a great potential to cover the whole food chain from production to processing, preservation, safety, packaging, transportation, storage and delivery. For instance, a variety of products exists such as nanopesticides, nanosized fertilizers, nanopromoters for plant growth, nanosensors, among others [19], which when applied cannot only come into contact with humans, but can actually be consumed along with the agricultural products. Due to their great potential to enter into the human systemic circulation system and interact with vital organs, carefully designed and comprehensive toxicity tests involving in vivo and in vitro have been carried out to assure safety to the human and environmental health.

Carbon-based NPs or carbon nanomaterials are a class of engineered nanomaterial with increased applications due to their exceptional optical, electrical, mechanical, and thermal properties. The individual NPs in this class include fullerenes, carbon dots, carbon nanotubes (CNTs), carbon nanobeads, carbon nanodiamonds, carbon nanofoils, carbon nanofoams, carbon nanofibers and graphenes [20]. Most of the carbon-based NPs have found wide application in agriculture, particularly as plant growth promoters and nanopesticides. For instance, due to their ability to effectively penetrate the seed coat and other plant tissues, carbon nanotubes (CNTs) have been used as plant growth stimulators [21]. The CNTs, single-walled carbon nanotubes (SWCNTs), have been shown to activate seed germination of corn, rice, switchgrass and tomato and enhanced the growth of different organs of corn, tomato, rice and soybean [22]. Similarly due to their superior electrical properties, CNTs have been extensively used for the development of biosensors. Thus, the NPs are usually functionalized or used in conjunction with other materials to minimize aggregation and enhance their usability. For instance, as reported in [21–23], surface functionalized CNTs were tailored with amino groups to control the efficient immobilization of acetylcholinesterase (AChE) onto the surface of glassy carbon electrode and enabled the construction of a highly sensitive organophosphorus pesticide biosensor in food stuffs where such pesticides were applied. In a bid to replace the agrochemicals, fertilizers and pesticides, carbon-based NPs have been used in the development of nanosystems for slow and controlled release of pesticides and fertilizers. As reported by [24], carbon nanofibers are used for making nanoparticles that contain pesticides and fertilizers specifically formulated to control their release into the seeds during germination.

Due to great potential for application in areas where these NPs can come into direct contact with humans, as shown above, the carbon-based NPs have comprehensively been studied on their toxicological impacts. The investigations into their toxicological effects, both for in vitro (on cell cultures) and in vivo (on organisms), have been conducted. Generally, these NPs have been observed to show some low or no toxicity in some cases [25], but in some other cases, however, the toxicity results have shown adverse effects. For instances, some studies conducted by Ostiguy et al. and Tao et al. (see [26, 27]) on some organisms using fullerenes, carbon nanotubes (single-walled or multiwalled) and nanofibers, have reported adverse effects. Similarly, a number of carbon-based NPs have also been shown to be cytotoxic to human alveolar epithelial (A549) cells, hepatocytes (Hep G2 cells), human embryonic kidney cells (HCT 116), and intestinal (P407 cells) cells [28]. Interestingly, while some studies conducted to investigate the effect of fullerenes and CNTs on plants have shown positive effects in terms of enhancing the plant growth and therefore could be commercialized as nanosystems for plant growth promoters; in other cases, a number of studies have shown that these NPs can have negative effects such as inhibitory effects against plant growth and against some beneficial microfauna [21]. Therefore to ensure that these NPs are safely used in a manner that human and environment health is ascertained, extensive and comprehensive risk assessments that include techniques that can capture delayed toxicity are required.

Inorganic-based NPs are probably the most diverse class of nanomaterials. NPs in this class include metals, metal oxide and quantum dots (QDs). They have unique chemical, physical, optical and quantum characteristics, and as a result, they have wide application in various fields such as medicines, engineering, environment and agriculture. In agriculture, as already stated, the use of huge amounts of pesticides and fertilizers results into serious environmental pollution with attendant adverse effects to humans, and sometimes, these agrochemicals affect the taste and nutritional quality of food crops. The advent of nanotechnology promises smart and intelligent nanosystems that can deliver the required nutrients to plants and nanoencapsulated slow release of fertilizers and pesticides that can deposit right doses at controlled rates. There are also nanosystems used as biosensors for detecting the presence of pesticides in agricultural products, which make use of inorganic NPs. While a lot of research is ongoing for development of such systems, already quite a few such systems are in use. For example, agribusiness and food corporations such as Monsanto, Syngenta, Kraft and BSF have already produced pesticides encapsulated with NPs [29]. Already NPs such as TiO_2, ZnO, MgO and a combination of other inorganic-based NPs, after being functionalized, have been utilized as effective nanopesticides [29]. The nanofertilizers are known to contain nanozinc, silica, iron and titanium dioxide, zinc cadmium selenide/zinc sulfide core shell QDs, indium phosphorus/zinc sulfide core shell QDs, manganese/zinc selenide QDs, gold nanorods, core shell QDs, specifically designed to control release [30]. Other inorganic NPs have been used as plant growth promoters, nanobiosensors among others. As reported by [23], ZnO NPs have been used as nanofertilizers and enhancement of nutrient absorption for plant growth. Within the wide context of agriculture, the quantum dots due to their characteristical electric and optical properties have been used as nanosensors and nanobiosensors. For instance, as reported by [31], cadmium selenide (CdTe) has been used as pesticide nanosensors for detection of 2,4-dichlorophenoxyacetic acid in food crops.

Because their application, particularly in agriculture, was envisaged to involve direct interaction with human biology and physiological systems through ingestion, these NPs have been widely investigated on their potential adverse effects. Toxicity of NPs and inorganic NPs in particular is well established. Literature is replete with cases where the inorganic-based NPs have been shown to cause both acute and chronic toxicity. The toxic effects have been observed in plants, animals, microflora and microfauna including cell lines. However, as observed by [32], most of the available data on ecotoxicology are limited to species used for regulatory purposes. That is, although the ecotoxicological data are available for aquatic and terrestrial organisms, it is predominantly from species deemed highly sensitive. For the purposes of understanding the toxicity potential of these NPs, these data are adequate. For cute toxicity, NPs such as Cu, CuO, Se, Zn and ZnO, and TiO_2 have been implicated in numerous studies [18]. For instance, silver and copper NPs were observed to cause adverse effects to both zebra fish and *Daphnia pulex* [33], while Cu NPs were observed to cause oxidative stress to earthworms, *Eisenia fetida* [34]. Commonly encountered metal oxide NPs such as CuO, ZnO TiO_2, SnO_2, CeO_2 and Fe_2O_3 have also been implicated in causing diverse adverse effects to organisms [28, 32, 34]. In terms of their use in agriculture, the long term or chronic effects of these NPs are of paramount importance. As reported elsewhere [35, 36], the quantum dots, metal and metal oxide NPs have been implicated in the long-term effects. Quantum dots are particularly toxic as they are usually made from already known toxic materials. For instance, cadmium-selenide (CdSe), cadmium-telluride (CdTe), selenide/zinc selenide (Se/Zn Se) and gallium (Ga) have been shown to cause immunotoxicity, oxidative stress and DNA damage [37, 38]. In most cases, the inorganic-based NPs are coated or encapsulated immediately after synthesis to prevent any aggregation and preserve their properties. Usually this surface functionalization can result into behavioral modifications, which in turn have a direct impact on their surface charge, size and reactivity. This then may be followed by a reduced toxicity.

The use of organic-based or polymeric/dendrimeric NPs in agriculture is equally wide spread. These NPs can be synthesized as nanowires or nanofibers and may be designed as hydrophilic or hydrophobic depending on the anticipated application. These NPs have useful characteristics that include biocompatibility and biodegradability, which confer upon them a multiplicity of application in various areas including agriculture. Examples of organic-based NPs include liposomes, vesicles, and micelles, dendrimers, nanocapsules and polymeric NPs. Like other classes of NPs, the organic NPs have equally been used in the formulations of smart-delivery nanosystems. For instance, the encapsulation of pesticides in the organic NPs ensures that there is slow and controlled release of the active ingredient, and therefore delivering more effective control over certain pests at lower dosage rates and over a prolonged period of time [39]. This reduces the overdosing and hence prevents pollution. Moreover, as smart systems, the nanopesticides are designed to increase the dispersion and wettability of agricultural formulations and unwanted pesticide movement. The nanopesticides have increased solubility and therefore can reduce contact of active ingredients with operators in the fields, thereby reducing the incidences of accidental toxic effects. Furthermore, these organic-based nanopesticides have the advantage of being biodegradable and therefore get assimilated into the soils, thereby adding some additional nutrients. Currently, there are quite a number of commercially available pesticides encapsulated by organic NPs. For example, [40], bifenthrin nanopesticide used for protection of agricultural crops has been formulated using polymers such as poly (acrylic acid)-b-poly (butylacrylate) (PAA-b-PBA), polyvinylpyrrolidone (PVP), and polyvinyl alcohol (PVOH). Similarly, fertilizers encapsulated with the organic-based NPs are commercially available and some of the commonly used organic materials include chitosan, nanocapsules (liposomes), polyethylene glycol (PEG), starch, cellulose, Poly(d, l-lactide co-glycolide) (PLGA) and polyester substances. Other smart nanosystems such as nanobiosensors for the detection of pesticides in food crop have been developed from the organic-based NPs.

As a result of envisaged wide application, the toxicological aspects of these NPs have been investigated. While some of the NPs have low or no established toxicity, some have been found to induce some toxic effects. For instance, polymeric, polyethylene glycol (PEG) NPs, Poly(d, l-lactide co-glycolide) (PLGA) NPs and solid lipid nanoparticles can cause immunotoxicity, nephrotoxicity and lung toxicity, respectively [41]. Similarly, some dendrimeric NPs such as polyamidoamine and poly (propyleneimine) have been investigated for their possible toxicity both in vitro and in vivo, and have been shown to have some concentration-based toxicity [42].

In general, the projected increase in the production and commercialization of NPs due to their novel properties will eventually lead to their increase in the environment with attendant increase in the exposure to organisms and hence with the concomitant adverse effects. But in terms of their use in agriculture, the kind of impact and adverse effects NPs may cause has probably not yet been clearly elucidated by the current risk assessment methods. A quick survey of literature shows that there has been an extensive evaluation of the toxic effects of NPs and currently these evaluations are still ongoing. It has been shown that most NPs exhibit some toxic effects, though in a number of cases conflicting results have been observed. There are numerous mechanisms by which different NPs induce their toxic effects, and these include cell proliferation, necrosis, apoptosis, DNA damage and oxidative stress among others. Interestingly, however, these same mechanisms have also been shown to be caused by environmental toxicants such as metal ions, pesticides, PCBs and other industrial chemicals [43, 44]. Thus, in investigating the minimal concentrations of NPs that can cause adverse effects, particularly in the actual environment that contain other toxicants, the risk assessment should involve aspects such as additivity, synergistic, potentiation and antagonistic effects. This kind of information is hugely beneficial in terms coming up with the regulatory framework and policies aimed at protecting human health with regard to nanotechnology in agriculture. This is because besides humans being exposed to NPs through nanotechnologically grown foods, they (humans) are also exposed to various other industrial chemicals through ingestion, inhalation and dermal contact. It is of no doubt, however, that the current assessment of the risks posed by NPs has a number of inherent limitations and uncertainties. The degree of uncertainty to a large extent is dependent on the application to which the NPs will be subjected.

3. Uncertainties of nanotechnology in agriculture

There is no doubt about the potential applications and benefits of nanotechnology in agriculture. In fact as research into the use of nanotechnology in agriculture matures, many more nanoproducts and nanosystems will be developed and commercialized to the benefit of the whole agricultural value chain. As observed earlier, the general risk in the application of nanotechnology in various fields has been reasonably assessed. In agriculture, however, the current risk assessment data do not seem to be sufficient for both industry and consumers to make informed choices about the use of this technology. This insufficiency of data leads to some significant uncertainties that relate to consumer and environmental safety, which is critical in the regulatory framework, and a necessary ingredient in giving public confidence in the products. In addressing the current state of uncertainties, there are a number of critical questions that need to be answered. For example, is current toxicity testing protocols sufficient to provide necessary information on delayed toxicity of NPs? Which dose metric best describes the toxicology of NPs, particularly through those that gain entry into humans through ingestion? Are there currently some validated techniques and methods that can detect the presence of NPs in the food matrix? Is there sufficient regulatory framework to ensure safety of NPs related to their use in agriculture? Is the NPs toxicity data from cell lines sufficient to inform regulatory framework? Are there some guidelines on the generation of NPs risk assessment data to ensure comparability of such data? Are risk assessment protocols used for both aquatic and terrestrial organisms sufficient to provide credible information for the exposure of humans through ingestion? What impacts will these nanosystems have on beneficial soil microorganisms? And finally, to what extent do these NPs accumulate and biotransform in plants? These and several other questions will be dealt with in this section as the issue of uncertainty in nanotechnology in agriculture is being discussed.

The question of whether the current toxicity testing protocols are sufficient to provide necessary information on delayed toxicity of NPs is one that speaks to the adequacy/inadequacy of the design of the risk assessments. The majority of the data from risk assessments is from traditional toxicity tests that rely predominantly on mortality and sublethal endpoints such as oral, dermal and ocular toxicity; immunotoxicity; genotoxicity; reproductive and developmental toxicity; teratogenotoxicity; carcinogenicity, growth, foraging, behavioral changes and among others. These toxicity tests are quite costly and time consuming [45]. Unfortunately, most of these tests do not necessarily capture the delayed toxicity and these do not give an opportunity for reliable prognosis about the ultimate effect on organisms. There is a suggestion that in order to understand the long-term impact that some of the NPs used in agrochemicals may have on human health and environment, more studies should begin to incorporate the genomics and proteomics techniques. These techniques though they involve the state of the art of instrumentation can prove to be faster and cost effective in the long term, particularly in the face of thousands of nanochemicals that are anticipated to be generated in the coming decades.

The aspect of the dose metric that best describes the toxicology of NPs, particularly, has been the subjective of debate among nano-ecotoxicologists for quite some time. Traditionally, mass has been used a dose metrics for most risk assessments for most NPs. However, other dose metrics such as surface area, number of particles, volume and size have also been investigated on their influence on toxicity, irrespective of chemical composition [46]. While in some cases, a particular dose metric could be responsible for the observed toxic effect, in other cases, another dose metric may be responsible. This creates some uncertainty, and thus, risks assessments for NPs need to ensure that all factors of a given NP type that lead to some toxic effects are clearly understood. This is particularly important because NPs have different characteristics. Thus, some NPs are soluble, while others are insoluble and further still some may be biopersistent.

Some uncertainty arises from lack of validated techniques and methods that can detect the presence of NPs in the food matrix. The detection and ultimate characterization of different types of NPs in agricultural food is necessary in understanding the benefits as well as the potential risks. Although some (few) methods for detection and characterization of such NPs are currently available, these methods need to be validated in addition to the need for the development of standard materials required in such methods [47]. Given that there are a number of NPs that are being developed for use in agriculture, need exists for research and development of more and validated methods required in the detection of NPs in agricultural products, especially food crops.

Although many countries are now setting definitions and regulatory frameworks for nanotechnology, the very nature of NPs in many ways makes it quite challenging in coming up with separate legislation that deals with these miniature materials away from their bulk counter parts. For example, as reported elsewhere [48], both the United States Food and Drug Administration (US FDA) and the United States Environmental Protection Agency (US EPA) have not recognized nanomaterials as the new chemicals and that nanomaterials do not require any new oversight. Ironically, these bodies (especially US FDA) require manufacturers of food products to demonstrate that the food ingredients and food products are not harmful to health; yet, as already stated, there is no regulation that specifically covers nanoparticles, which could become harmful only in nanosized applications. In the similar fashion, it is interesting to note that the European Union's main regulation covering nanotechnology applications is the REACH (EU Regulation on Registration, Evaluation, Authorization and Restriction of Chemicals) [23]. Generally, because nanotechnology is relatively new, at the global scale, there are currently no clear regulations governing the production, use, labeling and disposal of NPs/nanomaterials [21]. With the predicted increase in the production and commercialization of nanosystems for use in agriculture, there is a need for clear cut legislation and policies to protect and foster public health and confidence.

As already pointed out, the majority of the risk assessment data is from traditional in vivo animals tests. In as much as these tests can yield some useful information necessary to inform the regulatory framework, they are costly and time consuming. Additionally, the traditional tests normally involve one type of NPs at time. But humans will be exposed to these NPs used in agriculture together with other chemical contaminants. The data from these tests therefore may have an inherent degree of uncertainty. Recently, there has been some suggestion for using genomics, proteomic, transcriptomics, and metabolomics (the omics techniques) as high-throughput techniques, utilizing cell lines to cope with the rate of the production of the nanomaterials. Here again, there would be quite a number of uncertainties. For example, how reliable is the data from such techniques in terms of extrapolatability and predictability to human biology and physiology? Particularly, what is the degree of uncertainty for these data obtained from isolated cell lines kept in culture medium without the benefit of cross talk and interaction from other organs, as would be the case in the in vivo tests, have? The protective regulatory framework should always take into account the uncertainty to assure public confidence and trust.

There are quite a number of reasons why application of nanotechnology in the agriculture is still relatively at an infancy stage in comparison with other fields. The major ones include potential consumer health risks and a lack of unifying regulations and guidelines on risk assessment of nanotechnology. The use of nanotechnology in agriculture more than any application can lead to the introduction of NPs/nanomaterials into the human biology and physiology. Therefore, when risk assessment is not guided by unifying guidelines and regulations, then the risk quotient may be high and this can make the technology nonattractive to industry and consumers alike. In trying to harmonize the NP risk assessment data and ensure comparability, there is need for some guidelines on the generation of these data. Currently, one of the challenges relating to the usability of NP risk assessment data in regulatory framework is the somewhat conflicting nature of the toxicity results by different researchers. When there are specific guidelines to follow during the processes of conducting risk assessment of nanomaterials/NPs, the degree of uncertainty is minimized and regulatory framework can easily be formulated, particularly for a field such as agriculture.

Risk assessment protocols used for both aquatic and terrestrial organisms have contributed greatly in understanding the effect of NPs to organisms and to a large extent have provided credible information required for the development of safety guidelines on nanotechnology in general. However, with regard to application of nanotechnology to sensitive fields such as medicine and agriculture, new protocols and research designs of evaluating safety of NPs are required. For instances, are doses used in the actual environment, be aquatic or terrestrial, with a milieu of environmental matrices useful in extrapolating the effect to humans? And what is the contribution of other environmental toxicants to the observed toxic effects of NPs? These and several other questions need to be investigated in order to ensure nanotechnology safety in agriculture.

In addition to safeguarding human health as benefits of nanotechnology in agriculture increase, the safety of beneficial soil microorganisms which enable nutrient cycling and hence help to maintain basic soil fertility, need to be protected. Thus, there is need to carry out comprehensive NP risk assessment for all the NP types envisaged to be used in agriculture. And finally, more work needs to be done in investigating whether or not NPs can bioaccumulate and biotransform in plant materials.

4. Ethical concerns, public awareness and perceptions

Although nanotechnology is viewed as one of the key technologies of the twenty-first century and has major potential benefits, it has to be embraced with a precautionary measure, given that not much is known about its unintended effects on account of being new. Despite there being a lot of applications for nanotechnology in many fields and increasingly more applications are being employed in the field of agriculture, the general global population seems to know little about nanotechnology. Interestingly, however, as reported by [16, 49], a large proportion of the US and the European public have equally very limited knowledge about nanotechnology. These researchers concluded that despite the US public possessing little knowledge, a majority believed that benefits of nanotechnology outweigh the risks as compared with the majority of the European public who viewed nanotechnology with less optimism. It should be quite obvious that if public knowledge of nanotechnology in such highly advanced societies where literacy levels are relatively high is limited, then the situation is worse in other parts of world that are also grappling with high illiteracy levels. As expected, the level of knowledge of nanotechnology is much higher among the highly educated than those with least education.

When people know little about a technology, their perception and acceptability will to a large extent depend on how social and ethical issues concerning the technology are handled by industry and researchers. As stressed by [49], when knowledge is missing, people can use heuristics, such as trust, to assess the risks and benefits of a new technology. Thus, people are more likely to accept assurances about the safety of nanotechnologies, including nanotechnology in agriculture when they have higher levels of trust in the institutions, researchers and industry, emanating particularly legislative history. Another aspect that affects public perception about nanotechnology is the way the media reports issues on the technology. In less developed countries, the level of coverage of nanotechnology is very low and this is coupled with low levels of research in the technology. For developed countries with high levels of application of nanotechnology, the reporting or coverage of nanotechnology may be modest probably due to the specialized nature of the field and hence requiring specialist journalists who may be fewer [50].

Based on the factors that influences the perception and awareness of the nanotechnology in agriculture, there are quite a number of ethical issues that arise. In the face of the potential risks that nanotechnology in agriculture pose to human health and environment, should the industry continue using these nanosystems despite the uncertainty? Should there be a mandatory requirement for labeling of nanoenabled agricultural products, particularly food stuffs? Is it ethical that public/government institutions should continue funding development of nanotechnology products for use in agriculture despite the current levels of uncertainty? How much information should the public be made aware in relation to the nanotechnology in the whole agricultural value chain? Is it ethical for an industry to release nanosystems for use in agriculture to the public who have no idea about the potential negative impact on their health? Should there be regulations set in regulating nanotechnology in agriculture to increase public perception and acceptability? If these issues/questions are not fully addressed while the nanotechnology in agriculture is still in the development stage, the negative perception and hence reluctance of acceptance of this technology, similar to what was witnessed to genetically modified (GM) food stuffs, particularly in the European Union region may be experienced again.

5. Conclusions

There is no doubt that agrochemicals, fertilizers and pesticides have contributed greatly to the growth and increase in agricultural production. As observed, the last five decades has witnessed unprecedented increase in food production with only a marginal increase in cultivated land area. Despite huge benefits in terms increased agricultural productivity due to the agrochemicals, the excessive use of these chemicals has resulted into serious pollution to aquatic and terrestrial systems. The pollution has also resulted into increased disease burden, particularly to humans, as a result of consuming food and eaters contaminated with agrochemicals. The residues of pesticides have been detected and quantified in most agricultural food crops, while elevated levels of nitrates from chemical fertilizers have been found in both surface and ground water resources in various places across countries and continents. One of the main reasons for agrochemical pollution is due to yearly progressive increase in their application. For fertilizers, in some cases, only a small fraction of what is applied get utilized by plants. Therefore, the advent of smart nanosystems such as nanopesticides, nanofertilizers and nanobiosensors, among others, designed to increase solubility, enhances bioavailability and promotes targeted delivery and controlled release over a long period of time will immensely benefit the whole agricultural value chain. Thus, nanotechnology in agriculture will improve the efficient dosage of fertilizers, improve vector and pest management, reduce chemical pollution and ultimately decrease contact with agricultural operators.

The development of smart agrochemicals and other nanosystems for use in agriculture is still in the developmental stage. Of course currently, there smart nanopesticides, nanofertilizers and nanobiosensors that are in use and have made a huge impact in revolutionizing agriculture. However, the use of nanotechnology in agriculture has a number of risks, uncertainties and ethical concerns from the public perspectives. Different types of NPs that can potentially be used in the design and production of nano-agrosystems have been assessed in terms of their risk to human and environmental health. NPs from different NP classes such as carbon, inorganic and organic based have been subjected to safety evaluation. Interesting and useful data have been generated. However, the adequacy of the risk assessment for different NPs intended for use in agriculture remains an open question. Several issues have been raised about the sufficiency of the current risk assessment data for the formulation of protective legislation to human health from nanotechnology application in agriculture. Questions such as: are current toxicity testing protocols sufficient to provide necessary information on delayed toxicity of NPs? Which dose metric best describes the toxicology of NPs, particularly through those that gain entry into humans through ingestion? Are there currently some validated techniques and methods that can detect the presence of NPs in the food matrix? Is there sufficient regulatory framework to ensure safety of NPs related to their use in agriculture? Is the NPs toxicity data from cell lines sufficient to inform regulatory framework? Are there some guidelines on the generation of NP risk assessment data to ensure comparability of such data? Are risk assessment protocols used for both aquatic and terrestrial organisms sufficient to provide credible information for the exposure of humans through ingestion? What impacts will these nanosystems have on beneficial soil microorganisms? And finally, to what extent do these NPs accumulate and biotransform upon entry into plants? All these questions demand new approaches and perspectives in the design of risk assessment methods to ensure that humans and the environment are safeguarded from NPs potential harm as a result of their application in agriculture.

Other issues of concern that have been discussed about nanotechnology include low or limited knowledge of the general public about nanotechnology and low levels of publicity. Despite numerous benefits of any technology, when that technology is unknown, people will only resort to using heuristics such as trust to inform their perception about risks and benefits. If industry and the public regulators, for instance, FDA, have built a good relationship with the general public in terms providing good oversight, through trustworthy information, the public is inclined to believe when assured that a given product is safe. Furthermore, some ethical issues arise as to how much information the general public is given on the potential risks of the nanotechnology in agriculture. The role of the media is critical in shaping public opinion and perception about a given technology. Unfortunately, only few journalists are well schooled to report appropriately and effectively on issues of nanotechnology in agriculture. In order to gain public acceptance and avoid incidences of negative connotation of this new technology, similar to what was witnessed to genetically modified (GM) food stuffs, particularly in the European Union region, there should be adequate follow of information. The labeling of agricultural crops containing NPs should be encouraged to promote the free choice of use of such products by the public.

Author details

Phenny Mwaanga

Address all correspondence to: phennym@yahoo.co.uk

Department of Environmental Engineering, Copperbelt University, Kitwe, Zambia

References

[1] Pingali PL. Green revolution: Impacts, limits and path a head. Proccedings of the National Academy of Sciences of the United States of America. 2012;**109**(31):12302-12308

[2] Chauvin ND, Mulangu F, Porto G. Production and consumption trends in Sub Sahara Africa: Prospects for transformation of agricultural sector, United Nations Development Programme. Regional Bureau for Africa. Working Paper; 2012

[3] Tirado R, Englande AJ, Promakasikorn L, Novotny V. Use of Agrochemicals in Thailand and its Consequences for the Environment. Greenpeace Research Laboratories Technical Note 03/2008; 2008

[4] Mabe FN, Talabi K, Danso-Abbeam G. Awareness of health implications of agrochemical use: Effects on maize production in Ejura-Sekyedumase municipality, Ghana. Advances in Agriculture. 2017;**2017**:11. Article ID 7960964. https://doi.org/10.1155/2017/7960964

[5] Nonga HE, Mdegela RH, Lie E, Sandvik M, Skaare JU. Assessment of farming practices and uses of agrochemicals in Lake Manyara basin, Tanzania. African Journal of Agricultural Research. 2011;**6**(10):2216-2230. DOI: 10.5897/AJA R11.271

[6] Bhandari G. An overview of agrochemicals and their effects on environment in Nepal. Applied Ecology and Environmental Sciences. 2014;**2**(2):66-73

[7] Aktar W, Sengupta D, Chowdhury A. Impact of pesticides use in agriculture: Their benefits and hazards. Interdisciplinary Toxicology. 2009;**2**(1):1-12. DOI: 10.2478/v10102-009-0001-7

[8] Sunitha V, Reddy BM, Reddy MR. Groundwater contamination from agro-chemicals in irrigated environment: Field trials. Advances in Applied Science Research. 2012;**3**(5): 3382-3386

[9] Moss B. Water pollution by agriculture. Philosophical Transactions of the Royal Society B. 2008;**363**:659-666. DOI: 10.1098/rstb.2007.2176

[10] de Oliveira JL, Campos EVR, Bakshi M, Abhilashi PC, Fraceto LF. Applications of nanotechnology for encapsulation of botanical insecticides for sustainable agriculture and promises. Biotechnology Advances. 2014;**23**:1550-1561

[11] Nicolopoulou-stamati, Maipas S, Kotampasi C, Stomatis P, Hens L. Chemical pesticides and human health: The urgent need for a new concept in agriculture. Frontiers of Public Health. 2016;**4**:148. DOI: 10.3389/fpubh.2016.00148

[12] Aufan M, Rose J, Bottero J, Lowry GV, Jolivet J, Mark R, Wiesner MR. Towards a definition of inorganic nanoparticles from an environmental, health and safety perspective. Nature Nanotechnology. 2009;**4**:634-664

[13] Chellaram C, Murugaboopathi G, John AA, Sivakumar R, Ganesan S, Krithika S, Priya G. Significance of nanotechnology in food industry. APCBEE Procedia. 2014;**8**:109-113

[14] Joner EJ, Hartnik T, Amundsen CE. Environmental Fate and Ecotoxicity of Engineered Nanoparticles. Norwegian Pollution Control Authority Report No. TA-2304/2007. 2007

[15] The Royal Society and the Royal Academy of Engineering. Nanoscience and Nanotechnologies: Opportunities and Uncertainties [Internet]. 2004. Available from: http://www.nanotec.org.uk [Accessed: June 9, 2016]

[16] Siegrist M, Stampfli N, Kastenholz H, Keller C. Perceived risks and benefits of different nanotechnology tools and nanotechnology food packaging. Appetite. 2008;**51**:283-290

[17] Auffan M, Rose J, Orsiere T, De Meo M, Thill A, Zeyons O, Proux O, Masion A, Chaurand P, Spalla O, Botta A, Wiesner MR, Jean-Yves Bottero J. CeO_2 nanoparticles induce DNA damage towards human dermal fibroblasts *in vitro*. Nanotoxicology. 2009;**3**:161-171

[18] Bouwmeester H, Dekkers S, Noordam MY, Hagens WI, Bulder AS, de Heer C, ten Voorde SECG, Wijnhoven SWP, Hans JP, Marvin HJP, Sips AJAM. Review of health safety aspects of nanotechnologies in food production. Regulatory Toxicology and Pharmacology. 2009;**53**:52-62

[19] Rossi M, Cubadda F, Dini L, Terranova ML, Aureli F, Sorbo A, Passeri D. Scientific basis of nanotechnology, implications for the food sector and future trends. Trends in Food Science & Technology. 2014;**40**:127-148

[20] Mukherjee A, Majumdar S, Servin AD, Pagano L, Dhankher OP, White JC. Carbon nanomaterials in agriculture: A critical review. Frontiers in Plant Science. 2016;**7**:Article 172. DOI: 10.3389/fpls.2016.00172

[21] Grover M, Singh SR, Venkateswarlu B. Nanotechnology: Scope and limitations in agriculture. International Journal of Nanotechnology and Applications. 2012;**2**(1):10-38

[22] Jampilek J, Kralova K. Applications of nanotechnology in agriculture and food industry, its prospects and risks. Ecological Chemical Engineering Sciences. 2015;**22**(3):321-361. DOI: 10.1515/eces-2015-0018

[23] Parisi C, Vigani M, Rodríguez-Cerezo E. Agricultural nanotechnologies: What are the current possibilities? Nano Today. 2015;**10**:124-127

[24] Agrawal S, Rathore P. Nanotechnology pros and cons to agriculture: A review. International Journal of Current Microbiology and Applied Sciences. 2014;**3**(3):43-55

[25] Zakrzewska KE, Samluk A, Wierzbicki M, Jaworski S, Kutwin M, Sawosz E, Chwalibog A, Pijanowska DG, Pluta KD. Analysis of the cytotoxicity of carbon-based nanoparticles, diamond and graphite, in human glioblastoma and hepatoma cell lines. PLoS ONE. 2015; **10**(3):e0122579. https://doi.org/10.1371/journal.pone.0122579

[26] Ostiguy C, Roberge B, Woods C, Soucy B. Engineered Nanoparticles: Current Knowledge about OHS Risks and Prevention Measures. Studies and Research Projects, Report R-656. 2nd ed. Canada: Montréal; 2010

[27] Tao X, Fortner JD, Zhang B, He Y, Chen Y, Hughes JB. Effects of aqueous stable fullerene nanocrystals (nC60) on Daphnia magna: Evaluation of sub-lethal reproductive responses and accumulation. Chemosphere. 2009;**77**:1482-1487

[28] Perez S, Farre M, Barceló D. Analysis, behavior and ecotoxicity of carbon-based nanomaterials in the aquatic environment. Trends in Analytical Chemistry. 2009;**28**(6):820-832

[29] Chinnamuthu CR, Boopathi PM. Nanotechnology and agroecosystem. Madras Agricultural Journal. 2009;**96**(1-6):17-31

[30] Prasad R, Bhattacharyya A, Nguyen QD. Nanotechnology in sustainable agriculture: Recent developments, challenges, and perspectives. Frontiers of Microbiology. 2017;**8**: Article 1014. DOI: 10.3389/fmicb.2017.01014

[31] Duran N, Marcato PD. Nanobiotechnology perspectives. Role of nanotechnology in the food industry: A review. International Journal of Food Science and Technology. 2013;**48**:1127-1134

[32] Matranga V, Corsi I. Toxic effects of engineered nanoparticles in the marine environment: Model organisms and molecular approaches. Marine Environmental Research. 2012;**76**:32-40

[33] Griffitt RJ, Luo J, Gao J, Bonzongo JC, Barber DS. Effects of particle composition and species on toxicity of metallic nanomaterials in aquatic organisms. Environmental Toxicology and Chemistry. 2008;**27**:1972-1978

[34] Mwaanga P, Mbulwe S, Shumbula P, Nyirenda J. Investigating the toxicity of cu, CuO and ZnO nanoparticles on earthworms in urban soils. Journal of Pollution Effects and Control. 2017;**5**:195. DOI: 10.4176/2375-4397.1000195

[35] Casals E, Vazquez-Campos S, Bastus NG, Puntes V. Distribution and potential toxicity of engineered inorganic nanoparticles and carbon nanostructures in biological systems. Trends in Analytical Chemistry. 2008;**27**(8):672-683

[36] De Matteis V. Exposure to inorganic nanoparticles: Routes of entry, immune response, biodistribution and in vitro/in vivo toxicity evaluation. Toxics. 2017;**5**:29. DOI: 10.3390/toxics5040029

[37] Gagne F, Auclair J, Turcotte P, Fournier M, Gagnon C, Sauve S, Blaise C. Ecotoxicity of CdTe quantum dots to freshwater mussels: Impacts on immune system, oxidative stress and genotoxicity. Aquatic Toxicology. 2008;**86**:333-340

[38] Kim J, Park Y, Yoon TH, Yoon CS, Choi K. Phototoxicity of CdSe/ZnSe quantum dots with surface coatings of 3-mercaptopropionic acid or tri-n-octylphosphine oxide/gum Arabic in Daphnia magna under environmentally relevant UV-B light. Aquatic Toxicology. 2010;**97**:116-124

[39] Handford CE, Dean M, Spence M, Henchion M, Elliott CT, Campbell K. Awareness and attitudes towards the emerging use of nanotechnology in the agri-food sector. Food Control. 2015;**57**:24-34

[40] Khot LR, Sankaran S, Maja JM, Ehsani R, Schuster EW. Applications of nanomaterials in agricultural production and crop protection: A review. Crop Protection. 2012;**35**:64-70

[41] Yang Y, Qin Z, Zeng W, Yang T, Cao Y, Me C, Kuang Y. Toxicity assessment of nanoparticles in various systems and organs. Nanotechnology Review. 2017;**6**(3):279-289

[42] Duncan R, Izzo L. Dendrimer biocompatibility and toxicity. Advanced Drug Delivery Reviews. 2005;**57**:2215-2237

[43] Klaine SJ, Alvarez PJJ, Batley GE, Fernandes TF, Handy RD, Lyon DY, Mahendra S, McLaughlin MJ, Lead JR. Nanomaterials in the environment: Behavior, fate, bioavailability, and effects. Environmental Toxicology and Chemistry. 2008;**27**:1825-1851

[44] Barata CT, Varo I, Navarro JC, Arun S, Porte C. Antioxidant enzyme activities and lipid peroxidation in the freshwater cladoceran *Daphnia magna* exposed to redox cycling compounds. Comparative Biochemistry and Physiology, Part C. 2005;**140**:175-186

[45] Krewski D, Acosta D, Andersen M, Anderson H, Bailar JC, Boekelheide K, Brent R, Charnley G, Cheung VG, Green S, Kelsey KT, Kerkvliet NI, Li AA, McCray L, Meyer O, Patterson RD, Pennie W, Scala RA, Solomon GM, Stephens M, Yager J, Zeise L. Toxicity testing in the 21st century: A vision and strategy. Journal of Toxicology of Environmental Health B Critical Review. 2010;**13**(0):51-138. DOI: 0.1080/10937404.2010.483176

[46] Schmid O, Stoeger T. Surface area is the biologically most effective dose metric for acute nanoparticle toxicity in the lung. Journal of Aerosol Science. 2016;**99**:133-143

[47] Contado C. Nanomaterials in consumer products: A challenging analytical problem. Frontiers of Chemistry. 2015;**3**:48. DOI:10.3389/fchem.2015.00048

[48] Alfadul SM, Elneshwy AA. Use of nanotechnology in food processing and packaging and safety–review. African Journal of Food Agriculture Nutrition and Development. 2010;**10**(6):2719-2739

[49] Coles D, Frewer LJ. Nanotechnology applied to European food production: A review of ethical and regulatory issues. Trends in Food Science and Technology. 2013;**34**:32-43

[50] Dudo A, Choi D, Scheufele DA. Food nanotechnology in the news. Coverage patterns and thematic emphases during the last decade. Appetite. 2011;**56**:78-89